ERIC A. CHEEZUM

Chessie

A Cultural History of the Chesapeake Bay Sea Monster

JOHNS HOPKINS UNIVERSITY PRESS BALTIMORE

Printed in the United States of America on acid-free paper
2 4 6 8 9 7 5 3 1

Johns Hopkins University Press
2715 North Charles Street
Baltimore, Maryland 21218
www.press.jhu.edu

Library of Congress Cataloging-in-Publication Data is available.
A catalog record for this book is available from the British Library.

ISBN: 978-1-4214-4905-0 (paperback)
ISBN: 978-1-4214-4906-7 (ebook)

Special discounts are available for bulk purchases of this book. For more information, please contact Special Sales at specialsales@jh.edu.

For Ken

Who Always Believed

For Dan

Who Never Stopped Believing

For Margaret

Loyalest of Crewmembers

For Kelly and Brock

Who Have Kept My Ship Afloat

For My Parents

Stalwart Supporters of My Expedition

And for Violet

Who Missed the First Sighting

Can you pull in Leviathan with a fishhook, or tie down its tongue with a rope? Can you put a cord through its nose, or pierce its jaw with a hook? Will it keep begging you for mercy? Will it speak to you with gentle words? Will it make an agreement with you for you to take it as your slave for life? Can you make a pet of it like a bird, or put it on a leash for the young women in your house? Will traders barter for it? Will they divide it up among the merchants? Can you fill its hide with harpoons or its head with fishing spears? If you lay a hand on it, you will remember the struggle and never do it again! Any hope of subduing it is false; the mere sight of it is overpowering.

—JOB 41:1–9 (NIV)

CONTENTS

FIGURES

Chessie

INTRODUCTION

This book is about a monster, but not entirely the one you might be expecting.

A little over 45 years ago, in July 1978, people along the Virginia side of the Potomac River began reporting an odd *something* in the water. The anomaly, which eyewitnesses eventually labeled a "sea serpent," might well have gone unnoticed in the heat haze of an otherwise lazy summer. But 1978 had proved a fairly dull year for reporters up to that point, and the prognosis for the possibility of a newsy August was pretty grim. So, when word reached Tom Howard, then state news editor at the *Richmond Times-Dispatch*, that a mysterious aquatic creature had been sighted out in the provinces, it was manna from heaven. For a glorious few months in the summer and early fall of 1978, reporters chased the "serpent" up and down the banks of the Potomac, but the creature always remained out of reach, and it vanished as suddenly as it had first appeared.

Around 1980, sightings of the serpent—now dubbed "Chessie" by the press—began again, but this time they gradually drifted downriver to the Chesapeake Bay, and from there upward toward the northern end of the estuary, in the vicinity of Kent Island, Maryland. Whether the 1978 creature and the 1980 creature were the same thing is a matter for conjecture, but the public assumed they were, and that was good enough for an increasingly fascinated media. Described by eyewitnesses as "friendly" and unthreatening, Chessie quickly became a major pop culture sensation in Maryland and across the bay region. In 1982, a couple filmed what they thought was the serpent, beginning a serious scientific debate over the existence of what zoologists and video analysts now termed the "Chesapeake Bay Phenomenon." For a time, the serpent became the darling of cryptozoologists, scientists both amateur and professional who pursued evidence of cryptids, so-called hidden animals theretofore unknown to modern science. In the wake of this fame, Chessie nearly received protection under Maryland law and eventually became an icon for environmental advocacy. Although the creature never quite made it into the top tier of cryptids—the one occupied by, say, the Loch Ness Monster—by the time sightings petered out in the late 1980s, its prolific career had assured it a place in the mythology of the Chesapeake for years to come.

But—to ask a silly question—why? We might imagine that cryptozoologists expect to find sea monsters in every waterway, but why did everyday people living around the Chesapeake start seeing something unusual? The Chesapeake Bay is the largest estuary in the United States, with a surrounding watershed encompassing six states. For centuries, fishermen—watermen, as they are known locally—plied their trade in its waters without, as far as anybody knows, catching a single sea serpent in their nets or on their lines. What changed in the 1970s to cause Chessie to come to light and to become such a sensation in the next decade? What made the Chesapeake Bay sea serpent such an irresistible proposition?

One answer is that Chessie was an incredibly versatile metaphor. As though it were a swimming Rorschach test, people could read into the creature any meaning they desired. And, as we will see, there were *many* potential meanings. For cryptozoologists, for instance, Chessie was an icon of scientific advancement; for environmental advocates, the creature stood in for the complex ecology of the Chesapeake; for watermen, it was yet another example of urban gullibility; and for urbanites recreating along the shores of the bay, it represented nothing less than the power of the consumer to determine the purpose of water itself. It was poetic irony that Chessie shared its name with the holding company that owned the Chesapeake and Ohio Railway in the 1970s and 1980s: the monster certainly carried enough freight to be part of the Chessie System!

A Natural History of Sea Serpent Symbolism

If it seems strange for a sea serpent to inspire so many different perspectives, or indeed to be regarded as a metaphor at all, it is worth noting that Chessie and its eyewitnesses were carrying on a grand American tradition stretching back to the earliest days of the republic. Natural curiosities were a source of endless fascination during the Enlightenment, and their scientific study became inextricably linked with the young nation's development and expansion across the continent. For instance, Thomas Jefferson—the putative founder of North American paleontology—was fascinated by fossils of the mastodon and giant sloth and believed the animals still roamed the American West at the turn of the nineteenth century. Among the many other desiderata of the Lewis and Clark Expedition in 1804–5 was proof that these and potentially other strange creatures still existed. Jefferson's inquiries not only shaped scientific methods but also popularized science and natural history, linking the pursuit of both to national identity.[1]

The sea serpent was a special preoccupation of nineteenth-century natural historians, whose prestige might rise if they could obtain proof that such a creature existed. On this point, it is important to draw a distinction. Although the terms are often conflated and used interchangeably, there is a meaningful difference between a "sea serpent" and a "sea monster" that reflects the growing status of modern science as it emerged and developed during and after the Renaissance. Sea serpents were explainable and quantifiable: exaggerated specimens of known animals, perhaps, or merely products of misidentification. Whatever they were, they were susceptible, ultimately, to the authority of scientific observation. Famously, the "Great Sea Serpent" was one of the white elephants of Victorian science, especially after the crew of the HMS *Daedalus* allegedly spotted it near the Cape of Good Hope in 1848. A sea *monster*, by contrast, was akin to the biblical Leviathan or the Kraken of Scandinavian myth—evidence, if anything, of how far science had come in the effort to rationalize the natural world. Sea *serpents* became big news in the nineteenth century precisely because they reflected science's increasing domain, while sea *monsters* found themselves relegated increasingly to fiction and mythology. If the monsters still held any significance, it was to represent a primal fear of the unknown, of the boundaries of human knowledge—a perimeter naturalists were scarcely willing to admit existed.[2]

Probably the most famous sea serpent sightings in the antebellum United States took place near Gloucester, Massachusetts, in the 1810s. Following Jefferson's lead, American natural historians employed this serpent—unwisely, as it turned out—as a wedge to gain legitimacy among the European scientific establishment. The gambit failed, but the sea serpent lived on in the nineteenth-century American consciousness. Newspapers of the time are replete with often highly detailed and occasionally quite lurid accounts of encounters with the creatures, and the sea serpent became a pop culture shorthand to illustrate everything from political issues, to scientific discovery, to humanity's basic credulity. South Carolinian naturalist and sportsman William Elliott, for instance, capitalized on the association between New England and the Gloucester serpent to comment on abolition. Relating the tale of a "sea serpent hunt" near Charleston in the 1850s, he peppered it with barbs about this peculiarly "Yankee craft" sailing in Southern waters, secretly trafficking abolitionist propaganda into the slaveholding South. Along similar lines, an 1862 editorial posited that believing secessionists could be treated leniently and trusted not to rebel again was like believing in the sea serpent. "It is true no man of common sense ever believed in it, but semi-occasionally some old salt comes forward and makes affidavit to its precise length, its circumference and the number of humps upon its back, and all incredulous people must acknowledge that the weight of authority is against them," the paper sardonically opined.[3]

Natural science survived the cultural upheavals of the nineteenth century, but the sea serpent, at least as a subject of serious study, did not. Simultaneously the zenith and nadir of its study came in 1892 with the publication of *The Great Sea-Serpent: An Historical and Critical Treatise*, by the Dutch zoologist A. C. Oudemans. Oudemans theorized that the creatures in sea serpent sightings were probably a misidentified seal species (a proposition that reared its head again during the Chessie sightings).[4] Such speculation notwithstanding, the sea serpent/monster still retained some of its cachet at the turn of the twentieth century. The Progressives, in spite of, or perhaps because of, their scientism and expertism, still found such creatures useful as reference points for breakthroughs in natural science. As one Missouri newspaper wrote of a mosasaur fossil donated to the University of Iowa in 1900, "Once there were genuine sea serpents, or at least there were forms of reptile life that might have served every purpose of sea serpents could they have been perpetuated to the present time." "The Sea Serpent Is a Fact," proclaimed the headline of a *Washington Times* article in 1904. The paper had it on good authority—the Smithsonian Institution, no less—that most sightings of the creature could be written off as encounters with the elongated, serpentine oarfish, a deep-water specimen found only rarely near the surface. "Monsters of the Sea Now Recognized as Actualities Instead of Dreams of the Fictionist of Past or Present," the paper announced.[5] Twentieth-century sea serpents and monsters were to be purely functional: they existed to demonstrate the triumph of modern American science over old-world superstition. The press might report on encounters—and certainly *did*—but not without roundly debunking the stories at the same time.

With the coming of World War I and the threat of submarine warfare, sea serpents gained a new currency. The metaphor of a monster lurking under the surface of the water, just out of sight and waiting to strike, resonated strongly during the war, especially after the sinking of the *Lusitania* in May 1915. "The sea serpent was one of our most useful animals," a Wisconsin paper quipped in October 1915. "He provided conversation about the ocean that did not threaten international misunderstandings of any kind." Another paper summed up the sea serpent's fate more concisely, and in doggerel: "This festive snake bobs up no more, / Has quit the scene, / Quite likely put to flight before / The submarine." Before the war was over, German U-boats were being expressly compared to the sea serpents of yore.[6]

In the aftermath of the Great War, and on into the 1930s and beyond, sea serpents were gradually crowded out of the national imagination, although they continued to lurk on the edges of local and national debates about pretty much any issue. Even Prohibition, which might seem to us to be far removed from monsterdom, provoked the *Baltimore Sunpapers* to bring up the subject of sea serpents. "Prohibition may not prohibit," the paper observed wryly in 1921, "but then that leaves the disappearance of sea serpents from the summer resorts to be explained." On a similar theme, after the Loch Ness Monster became big news in 1933, and after Prohibition had been repealed late that same year, the *US Navy Review* responded that the "Scotch" creature "seems impressive, but let us see first what we can raise with the new domestic blends."[7]

During World War II the allure of sea serpents subsided. Although they once again doubled for the U-boat threat, the metaphor appeared less frequently than it had before. After the war, the decline continued. In the shadow of fascism and the Holocaust, on the one hand, and the atomic age, on the other, sea serpents were perhaps passé: too much like magic to be tolerated without demystification, and too old-fashioned to compete with the real monsters of the times. When sea serpents came up for discussion in the 1950s and 1960s, they appeared only with the understanding that they were a crude stand-in for the unknown, but not the unknowable—for the things that lay outside the totality of human knowledge at that moment, not forever.[8] With advanced space exploration and the moon landings just around the corner, Americans could no longer tolerate maps with areas marked "off-limits."

By the 1970s, the idea of a sea *serpent* had become almost quaint—hardly worth drawing the distinction anymore. In a decade when limits suddenly seemed to have been forcefully reimposed, *monsters* captured the American imagination, reflecting the uncertainty and disquiet of daily life. Like the terrifying great white shark in the 1975 film *Jaws*, monsters seemed to lurk in the most bucolic of places, unpredictable and unstoppable—like any number of threats that had surfaced throughout the era. Famously, even the beatific Jimmy Carter could not escape the clutches of a terrifying water monster, as he learned to his cost after a violent encounter with a swamp rabbit while boating in Georgia in 1979. "You will remember that the man from Plains met up with a ferocious aquatic bunny and fought for his life with a canoe paddle," the *Miami Herald* jeered afterward in an editorial titled "Cartership Down." To grapple with a sea monster could be empowering. Failing to take its measure, as Carter discovered, was crippling.[9]

It Comes from Inside

The fact that sea serpents had provided a handy metaphor for virtually any issue in American history does not completely explain what happened in the Chesapeake in the late 1970s and 1980s. Indeed, the gradual disappearance of the whimsical "sea serpent" across the twentieth century might suggest that the Chessie sensation was some kind of mass hysteria, or at least an episode of mass credulity. While that argument certainly came from some quarters at the time, the purpose of this book is not to disprove the creature's existence. Whatever it was that people were or were not seeing in the Chesapeake and its tributaries, they *thought* they were seeing something. With Chessie, *what* people saw is perhaps less important than *why* they saw it, and why it gained such notoriety in such a short span of time.

If Chessie symbolized anything in its earliest phase in the late 1970s, it was probably anxiety. There was certainly plenty of that to go around during the decade. Indeed, if a mad scientist were tasked with creating a monster from scratch purely by distilling the misery of a particular era, surely this is the one he would choose. Although it is true that Chessie first made the papers at a time when flashy headlines were thin on the ground, the news-cycle lull of August 1978 scarcely offset the ongoing parade of scandal, anxiety, and demoralization that characterized American life at the midpoint of the Carter administration. The first sightings of Chessie may have occurred in the bucolic setting of rural Virginia, but they played out against a cyclorama of recent national unrest that included political assassinations, stagflation, Watergate, the decline in public trust in the presidency and public institutions in general, defeat in Vietnam, and two energy crises. A year after Chessie popped up in the Potomac, Carter would complain on national television that the United States was suffering from a "crisis of confidence," and, looking back, it is hard to disagree.[10]

Americans responded to all these crises by rejecting traditional institutions altogether, and one way to do that was to embrace the occult and paranormal. Beginning in the late 1960s with the work *Chariots of the Gods?*, by Erich von Däniken, Americans became fascinated with the supernatural. Alien astronauts, pyramid power, the Bermuda Triangle, Bigfoot, the Loch Ness Monster—these were only a few of the subjects that gained popular attention during the decade. The pop culture zenith of this trend was the debut, in April 1977, of the television anthology *In Search Of . . .* For six seasons (the show ended in 1982), host and *Star Trek* actor Leonard Nimoy treated viewers to short documentaries focusing on mysteries, often of a paranormal nature. Traditional paths to knowledge, this movement suggested, could not be trusted. Science and its "experts" were fallible. Appearances were deceptive.

Given its focus on animals literally *hidden* from mainstream science, it should come as no surprise that cryptozoology exploded during this period. Before 1970, the number of titles dealing with the subject could be counted in the low dozens. By the end of the decade, publishers were churning out enough books to fill whole libraries.[11] Bigfoot and the Loch Ness Monster became internationally famous subjects of search expeditions, as enthusiasts sought to find legitimate evidence of cryptids in the wild. These enterprises were not without some slim expectations of success. Coelacanths, fish believed to have been extinct for over 60 million years, were famously discovered off the coast of Africa in 1938. Could it be possible that more prehistoric creatures otherwise unknown to science had survived into the present day, concealed from the view of modern society? More than half a dozen Loch Ness Monster hunts were undertaken in the 1960s and 1970s, with few tangible results, but the monster never lost its allure, and fifty years (and many debunkings) later, it still fuels public fascination. Dozens of similar creatures in North America and around the world, from the Lake Champlain Monster, to Tessie in Lake Tahoe, to Ogopogo of British Columbia's Okanagan Lake—and many, many more—have gone on to become household names in popular culture. Champ, the Lake Champlain Monster, is famous enough today to adorn one of the U-Haul company's moving vans. After at least a century of derision, the sea serpent/monster suddenly experienced a golden age, now signifying, at least for believers, the limits rather than the reach of science.

Existing on the boundary between the known and the unknown, cryptids also generated a fear more existential than mere angst. Operating from within a medium (water) that is both alien and potentially hostile to humans, the sea serpent/monster, with its random and ill-defined manifestations, is perhaps the most threatening cryptid of all. It thrives on liminality: breaking the surface of the water, violating the divide between land and sea, defying the ability to identify it properly. Without doing anything more than merely appearing, the sea serpent/monster unsettles, subverts, agitates. It summons a terror deep within the Jungian collective unconscious. As the German filmmaker Werner Herzog, producer of the 2004 mockumentary *Incident at Loch Ness*, told an interviewer in 2014, "What would an ocean be without a monster lurking in the dark? Like sleep without dreams." Religious studies scholar David J. Halperin puts the point more tersely. In *Intimate Alien: The Hidden Story of the UFO*, his provocative meditation on UFOs and the people who see them, Halperin observes, "*The UFOs came from within*, and they were seen, not because they were there—they weren't—but because they came bearing meaning."[12]

The sea serpent/monster comes from inside, too, and its meanings are legion.

Ring around the Bay

Yet myth and metaphor are not enough to explain this sea serpent/monster entirely. Chessie appeared suddenly, in a region where catching seafood—and knowing what is in the water to be caught—is one of the main industries, and where there was almost no previously recorded sea serpent activity, not even of the "silly season" variety. And the appearances *were* sudden, despite what many students of the monster, then and now, have claimed. The Chesapeake has never been much of a hot spot for cryptid activity—sea serpents least of all, ironically. Before Chessie, Maryland's best-known mythological creature was the "snallygaster," a half-reptile, half-chicken purported to haunt central and western Maryland. Chessie hunters seeking to establish a deep history for their serpent in the region inevitably cite a smattering of eyewitness accounts of varying detail stretching back into the nineteenth century. But there is no inherent connection between this handful of sightings and what appeared in the Potomac in 1978, and any continuity that has been inferred is the result of wishful thinking rather than reliable evidence. A sea serpent/monster should have been a tough sell in the Chesapeake. Why did this one catch on?

At the risk of demystifying Chessie too completely, there is a case to be made that it was a reflection of another, much more all-consuming monster: the mass migration of people onto the shores of the Chesapeake Bay. It is impossible to know for sure whether there was actually anything unusual inhabiting the waters of Virginia and Maryland after 1978, but one thing is certain: there were more people on the shoreline to see it. A quarter century after the end of World War II and the rise of mass tourism and suburbanization, the vast majority of communities situated within the "ring around the bay"—areas along the shorelines of the estuary and its tributaries—were still mainly rural, in terms of culture, economy, and population density. By the early 1970s, however, the tendrils of suburban development were reaching into those areas. In Virginia, sprawl gradually spread to the coastal communities of the Northern Neck from the urban centers of Richmond, Norfolk, and Washington, DC. The population in Westmoreland and Northumberland Counties, the two Northern Neck counties bordering the Potomac where serpent sightings proliferated, increased by 15.6 percent and 6.4 percent, respectively, across the 1970s. Queen Anne's County, Maryland, where Kent Island is located, increased a whopping 38.5 percent during the decade.[13]

The consequences of this movement of people were profound, although they took a few decades to become apparent. After World War II, the booming American economy spurred unparalleled affluence and explosive growth of automobile culture, mass tourism, and tract housing, developments that made the era a "golden age" for suburban growth and a sign of the vitality of the American dream.[14] By the 1960s, tourism officials in the mid-Atlantic, especially in Maryland, were actively identifying the "good life" with locales close to water. Accordingly, the industrial uses of the Chesapeake Bay were downplayed, or connected increasingly in a spiritual, rather than economic, way to the region. The region's famous seafood industry, for instance, gradually became a matter of "heritage" as much as productivity. The Chesapeake Bay was branded as a destination—a place in which to recreate and a state of mind to experience and commune with. By the time Chessie came along, the bay was no longer just *there*. It had a purpose and meaning, and was an end in itself.

Outsiders coming to the ring around the bay brought with them expectations for the water that differed from those of their native counterparts. For locals, water was primarily a place to work. Newcomers, by contrast, sought out the rural waterfront in order to have the Chesapeake "experience." Water, for them, was a place for recreation and aesthetic appreciation. The two positions were ultimately incompatible. For someone moving to the rural waterfront, or for the weekend tourist (who often moved there eventually), water had an almost mystical, ineffable quality. For those unused to the peculiarities of the water, an odd-looking wave pattern or half-glimpsed dorsal fin could easily be misconstrued as a strange animal of some kind—maybe even a sea serpent. Local fishermen might try to set the record straight, but they only *worked* on the water. What did they know? Suburbanites had a vested interest in insisting there was a monster in the water: Chessie's foreignness reflected their own, and validated their view of the water to boot.

That sense of foreignness, of not belonging, runs right through the first years of Chessie sightings and reflects a disquiet that was settling over American suburbia in the 1970s. As Kyle Riismandel has argued, "New, local threats undermined the expectations and understandings of suburban life as tranquil, safe, and family friendly; eroded the status of the suburb as a refuge from the conflict and discord often located in the city; and thereby created the neighborhood of fear."[15] By mid-decade, suburbs had become sites of ongoing conflict over environmental, social, and political issues—and of grassroots political organizing in response to those same issues. The Love Canal and Three Mile Island scandals rocked American suburbs just as Chessie was making its first appearances. Even in comparatively remote and pastoral places like the Northern Neck, how could one *not* feel unsettled? Riismandel's "neighborhood of fear" lived in a constant state of paranoia, terrified at the prospect of invasion by a legion of potential boogeymen. A sea serpent in the Potomac was just another monster threatening the Northern Neck's emerging suburban order.

At the same time, suburbanites and their suburbs were themselves invading rural America. They made for potent engines of social transformation, upsetting or erasing local traditions and social relationships, and leaving communities frayed and internally divided. Waterfront villages up and down the Chesapeake Bay and its tributaries inevitably became contested spaces as newcomers imposed urban, consumerist attitudes on the landscape and the native residents who lived and worked in it.[16] If the Chessie of the 1970s reflected the triumph of suburban attitudes in the Northern Neck, it also showed how ill-conceived some of those attitudes were.

When Chessie turned the corner into the 1980s, the decade of its heyday, there was a decisive shift in its meaning and potential. Partly, this shift came from the fact that Chessie migrated from Virginia, where even by 1978, suburbia was only just taking root in the Northern Neck, to Maryland, where development was more advanced. Accordingly, the Maryland iteration of the monster resonated differently. The work-versus-recreation and insiders-versus-outsiders themes faded, although they remained encoded in the creature's DNA and made occasional reappearances across the decade. In their place emerged a related issue peculiar to the 1980s: the privatization of formerly public spaces.[17] Beaches and marinas adjoining waterfront communities where sightings often took place were increasingly closed off to the general public, and local governments began regulating the water itself through noise ordinances and other laws. Chessie's fear factor vanished completely; by 1980, it was rebranded as a "friendly monster." Chessie was an insider in the new decade; it was a striking contrast to the days when the creature symbolized alienness.

Chessie's newly benign nature aligned congenially with another wave cresting in the 1980s: public celebration of the bay's heritage. Now Chessie seemed to embody the history, culture, and meaning of the Chesapeake—"Chesapeakiana." This development dovetailed especially well with tourism and marketing efforts initiated in Maryland in the 1960s, and it coincided with the rise of environmentalism in the Chesapeake, the explosion of modern media, and the emergence of the popular culture industry in the 1970s and 1980s. Through the monster, everyone could own a piece of the bay. In the 1980s, Chessie achieved its truest destiny: to become the Chesapeake's brand ambassador.

It was a fitting progression for an era when manufacturing and other blue-collar work was dying out—to be supplanted by the service and, eventually, information industries—and consumerism and materialism were in their ascendancy. Chessie's homegrown quality, even if it was not entirely authentic, neatly paralleled the Ronald Reagan aesthetic that emerged in the 1980s. Whereas Carter had done battle with the monsters of the 1970s and lost, Reagan would go on to tame them and to address those of the new decade—in both cases by applying his trademark homespun idealism and by unleashing the power of the local and individual.[18] Any monsters Reagan's United States might face would be *outside*: in the Soviet Union or, if they were in home territory at all, in the cities or on Wall Street. They definitely did not lurk in the heartland, source of America's greatest strengths.

When Chessie became national news in July 1982, columnist Robert McMorris of the *Omaha World-Herald* joked that Reagan ought to investigate the creature. "Maybe the Russians had something to do with it," he wrote. Even if the creature was not the vanguard of an invasion, searching for it still held merit: "It would be a welcome distraction for all of us, diverting attention from the cares of unemployment and inflation and West Beirut. We need that."[19] There was a grain of truth even in the jest: in Reagan's America, it was *just* possible that a humble sea serpent from the provinces really might be able to deliver the nation from its troubles.

Bay Monster

In the final analysis, if there is a real monster in this story, it is the Chesapeake Bay itself. It was the cauldron in which the disparate elements of tourism and suburbanization, environmentalism, consumerism, and a variety of other postwar trends came together and melded into an entity of surprising power and influence—a cultural leviathan that stalks every river and shoreline in Maryland and Virginia, and even beyond. Chessie the sea serpent is an avatar for the bay, and through its story, we can chart the parallel narrative of the bay's inexorable movement to the center of life in the mid-Atlantic.

This is why Chessie was such an irresistible proposition. If Chessie were a chimera or a mirage, then it merely reflected the noble but intangible nature of the bay's mystique. If Chessie were "real"—whether that means a snake, manatee, or some other, unknown animal—then it embodied the biological realities and responsibilities inherent in living on the shores of the nation's largest estuary. Chessie was simply a proxy: a focus for, and symptom of, public fascination. Its wake inevitably led back to the bay in the end.

Although this book explores many different themes relating to the Chesapeake, it is ultimately Chessie the sea serpent's story and is arranged accordingly. Chapter 1 charts the monster's initial appearances in the Potomac, as well as the conceptual clashes between locals and outsiders that provided the dynamo for Chessie sightings then and in the future. Chapters 2 and 3 follow Chessie into Maryland, where the creature gained fame and scientific legitimacy, becoming a full-blown cultural phenomenon in the state and region. Chapter 4 takes an excursion to Kent Island, Maryland, ground zero for Chessie sightings in the mid-1980s, on the one hand, and for suburban development, on the other. Here we see close up how the forces of postwar affluence propelled both Chessie and suburbanization.

The latter half of the book takes Chessie from the zenith of its popularity (or perhaps notoriety) through its gradual disappearance, seemingly, from Chesapeake waters. Much of this story is about Chessie as a symbol: chapter 5 sees the monster nearly receive protection under state law; chapter 6 explores its use by the US Fish and Wildlife Service as a mouthpiece for environmental education. In the final chapters, Chessie's story fragments. In chapter 7, the creature reaches its tenth birthday: "actual" sightings fall into decline, but the monster's grip on the region's imagination persists. In chapter 8, Chessie lends its name and mystique to a bull manatee discovered incongruously in the Chesapeake in the mid-1990s, an occurrence that ratified the monster sensation of the previous decade and made good on the role the creature had played in the environmental movement. Chapter 9 brings Chessie up to date, showing the enduring cultural power of Chesapeakiana and the monster's important role in embodying it. Cryptid or estuary, there will always be a monster haunting the Chesapeake.

In a book of this nature, it is tempting to try to "explain" what Chessie was or was not. Such an enterprise is an exercise in futility. As we shall see, there was no consensus on that subject at the time, and hindsight, while offering a few strong possibilities, nonetheless still cannot provide a conclusive answer. More important, it violates the spirit of Chessie, which was (and still is) a subject that rightly evoked wonder and affection, especially among kids of the era, who thrilled at the possibility that the Loch Ness Monster lived in their backyard.

Ultimately, the purpose of this book is not to pass judgment on the monster's existence or to cast doubt on the many eyewitnesses who reported sightings in good faith and sincerely believed in what they were seeing. Instead, we embark now on an expedition to find the *meaning* of the creature and to determine what that meaning tells us about the times and places Chessie swam through. As we shall see, this sea serpent was every bit the "Chesapeake Bay Phenomenon" its fans believed it to be—just not, perhaps, in quite the way they imagined.

Chapter 1

Catching Monster Fever

ONE EVENING IN LATE JULY 1978, retiree Donald Kyker was relaxing on his back porch, looking out at the Potomac River, when he noticed something odd out in the water. A recent arrival at the Bay Quarter Shores retirement community located near Heathsville, in Virginia's Northern Neck region and near the mouth of the Potomac, Kyker was a former employee of the CIA and possessed what he called a "strong power of observation." What he saw defied his analysis. News reports nearly a month later described the object as "the size of a telephone pole with its head raised above the water, moving down the river about 60 yards from shore. . . . Instead of undulating sideways, it undulated up and down." "It was absolutely smooth, smooth as a snake or an eel," Kyker was quoted as saying, "but it did not move through the water like a snake or an eel."[1] At first, Kyker believed he was witnessing a practical joke of some kind, or perhaps a group of otters swimming in a chain, but he quickly rejected these possibilities after neighbors downriver claimed also to have seen the creature. When newspapers reported the incident, experts on marine life in the region were quick to dismiss the sighting as a misidentification by inexpert eyes. But Kyker stood his ground—and the Chessie sensation was born.

Kyker's neighbors, Howard and Myrtle Smoot, who owned a vacation home downstream in the same development, reported *four* strange creatures in the river, two of which they estimated at 30 feet long, and two at about 15 feet. Howard Smoot immediately went for his rifle and shot one of the creatures in the neck. It was not clear if the creature was injured: "It rose out of the water and disappeared," Myrtle Smoot told reporter Anne Hazard of the *Richmond Times-Dispatch*. "Having children and grandchildren who swim and ski, we wanted to find out what it was. We didn't know whether it was dangerous, and we figured if we could get one, we could find out what it was." After firing on the creatures, members of the Smoot family took a rowboat out to where they had been in the river, to no avail. C. Phillip Stemmer, the Smoots' next-door neighbor, was with them when they saw the creature. "Everyone who saw this was stone-cold sober as far as I know," he told Hazard. "It was the oddest thing I've ever seen. I can't relate it to anything."[2]

Whatever it was that Kyker, the Smoots, and Stemmer saw, the press fell in love with it instantly—probably because talk of sea monsters in the Potomac sold newspapers during the summer news doldrums. Within a couple of days of the Kyker report, the *Times-Dispatch* had identified some 20 people who claimed to have seen the serpent at various times, and the paper reported all manner of serpentine news, however tangentially related it might have been to the current situation. The story was distributed to the Associated Press, and the Potomac River monster briefly became a national sensation. "Everybody knows Britain is having its economic problems," the *Richmond News Leader* quipped the day after Kyker's sighting made headlines, "but

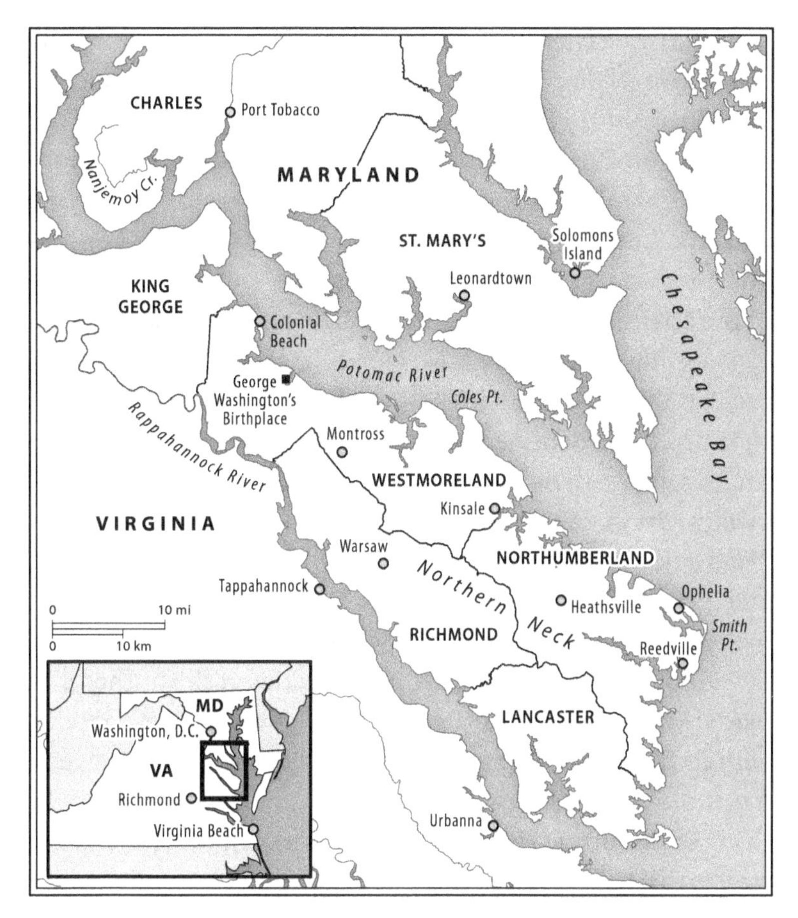

Virginia's Northern Neck region.
ERIN GREB CARTOGRAPHY

when the Loch Ness Monster deserts, things have gone too far. Our essential hope is that the beast has sufficient sense to settle in the Old Dominion."[3]

The Anaconda Syndrome

Even if the creature lacked good sense, it was certainly *sensational* enough to give it staying power. News of the monster's appearance electrified the public, even in the staid and slightly remote rural communities of the Northern Neck, plugging the area directly into the mains of 1970s paranoia. Not all of the current, however, flowed in the creature's direction. Contention quickly developed between eyewitnesses and officials, with the latter camp brushing aside the possibility of an unknown animal in the river with clinical ease. The first member of the scientific community to publicly pass judgment on the Kyker sighting was Robert M. Norris Jr., executive secretary of the Potomac River Fisheries Commission, who suggested that Kyker had misidentified a pod of porpoises. Kyker had made the mistake, Norris implied, because porpoises were an unusual sight in the Potomac. The proposition irritated Kyker. "Had it been a school of porpoises, it would have had dorsal fins, and it would have had a diameter larger than seven or eight inches," he tersely retorted to reporter Hazard.[4]

The first shots had been fired in a series of skirmishes that played out in the papers over the course of the next few months, with each side claiming greater knowledge than the other when it came to the water, what belonged in it, and who had the right to say so. As further reports of the monster came in, the experts started to pay closer attention, although none were willing to admit the possibility of a cryptid in the river. The serpent Kyker described, if it were real, was "one hell of an animal to be in the bay," John V. Merriner, chair of the Virginia Institute of Marine Science ichthyology department, diplomatically told Hazard. James E. Douglas, chair of the Virginia Marine Resources Commission, was less equivocal. "It's nothing more than a big snake," he told the papers flatly.[5]

Officials were not basing their statements purely on their credentials and personal experience; they were reiterating what seemed an obvious truth drawn from the experience of local people who worked on the Potomac—watermen, as they were known locally. As Virginia Institute of Marine Science official

Donald Kyker's sketch of the creature he saw in the Potomac in July 1978.
COURTESY OF LINDSAY LEAVITT AND THE KYKER FAMILY

John A. Musick bluntly put it, "Fishermen who spend much of their time on the water 'would have seen something if there was anything to see.'"[6] It was only natural for officials to appeal to the knowledge and experience of the waterman, who encountered the Potomac and its tributaries on a daily basis, and with whom conservation authorities and the scientific community had an intimate and symbiotic relationship.

In other times, the discussion might have ended there, along with the monster's nascent career. But in the "me" decade, truth was a more flexible concept than it had been in the past and often turned on the matter of what was *not* being said rather than what *was*. We have no way of knowing what watermen were saying about the creature in private conversations among themselves or with authorities, but they certainly were not discussing the matter with reporters. Their reticence did not, therefore, close the debate as officials assumed it would. Instead, in a trend that would follow the creature for the duration of its travels, it encouraged believers to imagine that watermen had an ulterior motive of some kind. Perhaps they were not speaking out in fear of their reputations. Perhaps they were part of a cover-up. Whatever the reason, maybe watermen were *not* best equipped after all to identify a sea serpent.

With a definitive explanation for the sightings unavailable, and the route to discovering one already becoming epistemologically murky, other explanations filled the void. Naturally, the press went for the obvious, making the obligatory link to the Loch Ness Monster and other notable lake monster phenomena popular in the era.[7] It was the proposition that the monster was a "big snake," however, that best fit eyewitness descriptions, and it was that explanation that stuck—probably to the dismay of James Douglas, who had uttered just those words. Douglas had no doubt been referring to an unusually large example of a known species of water snake, but like the creature itself, the words became exaggerated. Within days of the first sighting, the *Times-Dispatch* had already run an article about Lewis R. Bray, a diesel mechanic who claimed he had encountered a giant serpent while he had been tilling a field across a creek from the Bay Quarter Shores retirement community—in the 1940s. According to Bray, the snake "moved by undulating vertically like a caterpillar, instead of sideways like a snake." He attempted to kill the snake by running it over with his farm implement, but the blades on the machine had no effect, and the serpent eventually tried to strike at him. Terrified, Bray drove off and never saw the animal again. The anecdote was definitely the stuff of nightmares, and reporter Hazard only added fuel to the fire when she proposed that Bray's encounter—and, by implication, the Kyker creature—might connect to local lore about a giant "land serpent" that had allegedly existed in the area.[8]

The land serpent story reared its head in a more official capacity during an August 18 meeting of the Virginia Marine Resources Commission. There, a Maryland waterman named James W. Dutton came forward and claimed that he had seen the *track* left by a large snake in the vicinity of Nanjemoy Creek, a tributary on the Maryland side of the Potomac. Dutton voiced what ultimately became the media's favorite explanation for the monster: large snakes, he said, had come to the Potomac aboard South American ships that were abandoned on Nanjemoy Creek many decades before. Their ophidian cargo escaped into the creek and surrounding marshes, and from there into the Potomac. Oddly enough, there really *are* accounts of giant snakes menacing Marylanders in the later nineteenth century, although it is doubtful Dutton had ever heard of them, and in any event, they did not take place near Nanjemoy Creek.[9] And as for those South American ships, when *exactly* were they supposed to have landed in southern Maryland? The seventeenth century? Eighteenth? Nineteenth? No doubt Dutton was sincere, but his theory was light on detail and did not hold up to much scrutiny.

No matter: the press adopted the land serpent story and ran with it, dubbing it the "Anaconda Syndrome." The new name alone carried the thrill of a B movie, riffing as it did on the "China Syndrome," a popular term for cataclysmic nuclear meltdowns that was used the next year as the title for a movie on that theme. It also evoked ecological disaster, the idea of "nature gone wild"—a terror that seemed all too possible in the 1970s, and that had been brought viscerally to life in *Jaws* (1975). As Heathsville resident Elizabeth Hinton put it, "I just finished reading *Jaws*. That shark didn't belong in those waters, so I hope we don't have anything around here that doesn't belong in this part of the country."[10] Even closer to the knuckle, perhaps, was *Piranha*. Released nationally only weeks before the first monster sighting was reported (although it was apparently not screened in Northern Neck theaters until early autumn), that film featured a school of scientifically augmented carnivorous South American fish invading and terrorizing the waters of suburban America. Whatever the Anaconda Syndrome did, or did not, boast in plausibility, it successfully melded homegrown folklore with 1970s suburban paranoia in a way that made sense to the public. For the emerging suburbia of the Northern Neck, the creature was a NIMBY by proxy: perhaps not as realistically terrifying a prospect as nuclear meltdown or polluted groundwater, but a nebulous threat that residents wanted to exorcise from their backyards just as urgently.[11]

Despite its lurid and anecdotal nature (or maybe because of it), the Anaconda Syndrome became the media's go-to explanation for the monster, long after it had left the waters of the Potomac. One reason for its allure was that it reflected genuine fears of invasive species in the Chesapeake region, and of environmental decline in general. Although the environmental movement would not fully reach the Chesapeake until the 1980s, the 1970s nonetheless was a period of growing concern, especially at the state level, over the health of the bay and its watershed. One of the icons of the peril the Chesapeake faced was the nutria, a beaver-like rodent native to South America that was first brought to the Delmarva Peninsula in 1943. Two factors made the nutria particularly dangerous to the Chesapeake ecosystem: its diet, which consists mainly of wetland vegetation, and its incredible ability to reproduce. By the 1970s, marshes on the Eastern Shore of Maryland were infested with nutria colonies, and southeastern Virginia had likewise been invaded, probably by nutria migrating northward from North Carolina. Although the nutria has not, to date, been found in the Northern Neck, it was a looming possibility in the 1970s, even more so than a South American anaconda.[12]

The Anaconda Syndrome also unleashed a more general, existential fear on residents of the Northern Neck, where, as the *Times-Dispatch* recognized, "the economy is tied to the water, and to farms and timber tracts along the water." Nearly every aspect of life in this rural area was connected in some way to otherwise familiar waterways that could now be harboring a potentially dangerous alien creature. Interviewees who had radically different relationships with water all expressed trepidation about it now that a monster could be lurking there. As one person put it, "Seafood processors, watermen and resort developers 'get their livelihood from the water, and it's strange to think there's something like this around.'"[13]

In spite of the concern expressed for people working on the water, however, most of the alarm about the monster came from people who, at least as far as the papers reported, engaged with it through some form of leisure. For instance, Myrtle Smoot, who had been part of the initial sighting at Bay Quarter Shores, told the papers that she was reluctant to allow her son to go water-skiing for fear of the creature. Another woman, Ophelia summer resident Virginia Ingram, agreed, speculating that "this is keeping people out of the water as they hear about it."[14] A couple who owned an excursion boat operation based in Reedville reported that their passengers were "looking for the serpent." Judy L. Brincefield, who claimed to have seen the creature early one morning while visiting the riverfront restaurant of the Northumberland Plantation recreation development, remarked that foot traffic had increased on the beach since sightings had started. "I don't know whether it's because the season's about to end or what. People are talking about it, that's for sure."[15] Well, *some* people; watermen certainly were not going out of their way to talk to reporters. The distinction was significant and would only keep growing as the supposed serpent swam on.

By the latter third of August 1978, the monster sensation had reached a crescendo. For the next few weeks, sighting reports came out of the woodwork, from people from all walks of life and from locales stretching down the Potomac, and even as far away as the mouth of the Chesapeake Bay. A company foreman in Ophelia; a tugboat and yacht captain working in the Rappahannock River; a grain manager for the Southern States Cooperative in Urbanna—all professed to have seen something strange and serpentlike in the waters at various points along the Potomac. One report even came in from as far afield as Virginia Beach.[16] Marine experts tried in vain to discount the sightings, but witnesses refused to concede. "These things start up in the newspaper, and then everyone sees one," an exasperated Musick complained to the *Times-Dispatch*. "It's like *Jaws*."[17]

The diversity of sightings led to an equally diverse range of "explanations" for what people were seeing that ran parallel to the Anaconda Syndrome. Most of the possibilities that were floated were serious attempts to find a "natural" answer to what the creature was, but a few were pretty outlandish. On the serious side, there was the idea, put forward shortly after the Kyker sighting, that the monster was really a pod of porpoises. There were many variations on this theme involving different animals, including seals, stingrays, families of otters, and a manatee. There was also the possibility that the creature was really an inanimate object, such as a fishing float or log, that was either untethered and moving with the current or anchored in place and simply misidentified. In the less likely column, one suggestion was the ribbonfish, although these are deepwater fish unlikely to survive in the brackish waters of the Chesapeake, and they bear only superficial similarity to the typical monster description. And on the bizarre end of the spectrum, one man who claimed to see the monster believed he had glimpsed a submarine on maneuvers out of Norfolk; another jokingly suggested the creature was really drug traffickers trailing bags of marijuana behind a powerboat.[18]

Seemingly powerless to influence the debate along scientific lines, officials at least tried to stress that the "creature," whatever it was, was almost certainly part of the local ecosystem, that it was not dangerous, and that people should not try to harm it. Marine Resources commissioner Douglas particularly objected to Howard Smoot's attempt to shoot the creature and had even entertained introducing a resolution before the Potomac River Fisheries Commission declaring it "friendly," in an effort to prevent such acts in the future.[19] But as the list of potential explanations shows, the public had a pretty high tolerance—an *expectation* even—for strange objects and animals swimming in the

Potomac. The reality of this, whether or not anybody thought about it at the time, could not help but undermine officials' efforts to normalize the thing people were seeing and calm down the situation. It also tells us a little about the cognitive dissonance of the modern world. For all that residents of the Northern Neck fretted about giant, killer snakes potentially lurking in the creeks and along the byways of their Virginian Brigadoon, no one thought twice about drug dealers (joke or otherwise) or a nuclear submarine doing the same thing. If a sea serpent could not be a normal part of the Potomac River ecosystem, why could they?

Big News, Little Sea Serpent

About the same time the media was conjuring images of anacondas living along the river, the name "Chessie" was first attached to the monster. Although Anne Hazard reported on the monster for the *Richmond Times-Dispatch*, her articles were edited for national distribution by the Associated Press. Writing in 1994, *Times-Dispatch* reporter Tom Howard remembered, "Syd Courson, an Associated Press rewrite editor in Richmond at that time, gets the credit for naming the monster 'Chessie.' He took the *T-D* articles nightly and rewrote them into AP style. At one point, he made up the name 'Chessie' and worked it into one of the articles. He told me later it just seemed to fit."[20] It is unclear exactly when the name first appeared in print locally, but references to "Chessie" exist in papers carrying the stories nationally as early as August 18, 1978.[21] The nickname took a while to stick, however. Hazard appears *never* to have adopted the nickname in the articles she wrote in 1978, and even Courson seems only to have used it in stories rewritten for use outside Virginia. The name did not become widely used until 1980.

The serpent seems to have fired Courson's imagination. Besides naming it Chessie, he also used it for the first time as a vehicle for political commentary—a practice that became routine in later years. Published in September 1978, Courson's satirical mock interview with Potomac "Potty" MacTavish attempted to use the monster's status as a sort of outsider to make pointed observations about 1970s activism and other hot-button issues. So Potty announced a "new wave of militancy by sea serpents" and "told" Courson that he and other sea serpents in the Potomac were "thinking about filing a class-action suit on the basis of our being a minority and all that. . . . It's worked for blacks and women, why not us?" He also suggested that since he and his family had lived in the Potomac for three centuries, they might claim the Chesapeake as their own territory, just like "those Indians up in Maine." Here Potty was referencing two major court cases that had recently been in the news: *Regents of the University of California v. Bakke* (1978), a Supreme Court decision that upheld, but blunted, affirmative action; and *Joint Tribal Council of the Passamaquoddy Tribe v. Morton* (1975), a US First Circuit Court of Appeals decision that enabled two Indian tribes to make land claims against Maine based on aboriginal title. Potty also went on to take shots at the Jimmy Carter administration's foreign policy and even suggested he might become an antipollution activist.[22] It was not subtle commentary, to be sure, but it nonetheless demonstrated the monster's flexibility as a piece of pop culture. It also pointed the way ahead: this would not be the last time that the monster would weigh in on public issues—especially environmental ones.

As awareness of the monster bloomed, its potential as a metaphor grew too. A pair of related *New York Times* entries from late 1978 show the disparate, but connected, ways that the monster enabled commentators to talk about different aspects of popular culture. Writing in November, Richard D. Lyons played up the "comic overtones that accompany a suggestion that a strange creature inhabited the Chesapeake." In his estimation, whatever people like Kyker had seen was either "a monster or a joke, depending on one's point of view"—and there was little question where Lyons stood on the matter. Peering down at the scene in the Northern Neck from his vantage point in faraway New York City, Lyons could not avoid adopting a Menckenian tone toward the antics of rural America. His article concluded by juxtaposing two images: a skeptical marine biologist's suggestion that the monster was really rays causing the water to undulate and boil, and "another 'thing' . . . spotted in the local waters lately." This latter "thing" was a dummy monster built to "liven up" a fish fry in Kinsale, Virginia. Lyons described its fate as though the model had been built in effigy in order to exorcise some kind of demon: "They thought they had found a way to destroy their monster once and for all. They shot it with a toy cannon. But weeks later, Myrtle Smoot's 11-year-old son spotted the creature for the second time, and the monster story started all over again."[23]

In a good-natured rebuttal to Lyons in December, *Times* reader William Worthington responded that "maritime Yankees, particularly those from the littoral of Massachusetts Bay, readily believe that Chessie, the mysterious being seen in Chesapeake Bay, is not a monster, nor a joke. Probably she is a granddaughter of Samantha, who was perhaps the best known of the clan." Worthington was referring to the titular hero of Stephen Vincent Benét's short story "Daniel Webster and the Sea Serpent" (1937), in which then-secretary of state Webster enlists the aid of "Samanthy," a New England sea serpent, to intimidate the British into concluding the Webster-Ashburton Treaty of 1842. Thinking that the United States has enlisted the biblical Leviathan into its navy, British lord Ashburton realizes that his country's naval supremacy may not be as secure as he had believed. Having earlier refused to negotiate on some of the points in the treaty, Ashburton relents, and for her help, Samanthy is commissioned into the navy and ordered to find a mate and build a fleet in the Pacific. There was a sting in the end of Worthington's summary of Benét's tall tale: "And now at last a member of this self-perpetuating squadron has returned to the Potomac. One wonders why. Perhaps for recommissioning; perhaps, indeed, to serve Secretary Vance as her ancestor had aided his predecessor. But, please, not to be discarded, junked, or destroyed."[24]

The reference to Cyrus Vance, Carter's secretary of state from 1977 to 1980, drives home yet again how potent the Potomac monster could be. A swimming enigma, entirely defined by its observer, the monster was an empty vessel capable of expressing virtually any viewpoint or emotion. In just a few short months, the monster had become an expression of fear of the unknown, worries about pollution, conflict between science and popular knowledge, regional and national pride, and more. Now the monster might have swum home to rescue the United States from the vacillating foreign policy of the Carter administration. Were there any waters this monster could not inhabit?

Ironically, just as the creature had begun to flex its muscles as a pop culture phenomenon in the fall of 1978, actual sightings petered out. None were reported in 1979—but a lack of fresh material did not put off the press. The *Times-Dispatch* ran two separate retrospectives about the monster that year, in January and November. The stories revisited the experiences of Kyker and the Smoots, as well as those of other interviewees—but perhaps with a more skeptical eye. With no further sighting reports or photographic proof, the plausibility of the whole phenomenon was up for debate. As Hazard's headline glumly concluded in January 1979, "'Monster Fever' Has Cooled."[25]

Bless Goodwin Muse

For over a year and a half, the Northern Neck was bereft of sea serpent manifestations. It was an exceedingly long period in news cycle terms, even for the 1970s, and in that time the press had largely moved on from Chessie. So it must have come as a shock to everyone involved when the *Times-Dispatch* announced that a new sighting had taken place on June 14, 1980. "'It Looked like a Snake,' Potomac Farmer Says," the paper's headline gasped. "A farmer who has lived all his 59 years on the shores of the Potomac River says he, his wife and four friends saw a sea monster Saturday," wrote reporter Brad Cavedo. The farmer in question was one Goodwin Muse, whose homestead was located on the Potomac River waterfront, adjacent to the birthplace of George Washington. Muse and the other witnesses gave a description similar to those reported in 1978: snakelike in form, but unusually long at 10 to 14 feet, about 5 inches in diameter, with a fist-sized head that projected out of the water as it swam. What Muse did *not* see, he took pains to explain, was the vertical undulation typical in accounts of two years before—a hitch in the ongoing construction of Chessie lore that would be quickly smudged over.[26]

Even though Muse's sighting did not quite fit the mold, newspapers relied on it because Muse brought an important component to the story that had been missing when the monster first appeared: *himself*. Almost to a man, the monster sighters interviewed in 1978 had been retirees, vacationers, or seasonal residents of the Northern Neck. Reporters following the monster story then had lamented the lack of an eyewitness account made by an indisputable waterman, who could imbue news stories not only with vocational knowledge of the Potomac and its wildlife but also with a little local flavor. Indeed, although the papers had consulted numerous marine biologists and officials with practical and scientific experience of the Chesapeake watershed and its fauna, none of their opinions about the potential for the existence of a river monster—all in the negative, of course—ever gained much traction with the public. The holy grail of sea monster eyewitnesses was a local waterman willing to go on the record or, failing that, a local person of good standing in the community. Muse, although not a waterman, was the next best thing. Not only was he credible, he was a *native*. "I've been here all my life around this river," he was quoted as saying in the *New York Times*, "and this is something I've never seen before."[27] The experts took him at his word. "I have complete faith in anything Goodwin Muse would say. He's a sober, industrious person who's lived on the water all his life," Robert Norris told reporters in September 1980.[28]

Another sighting quickly followed Muse's, on June 22. That Sunday, G. F. "Buddy" Green III, the district manager of a finance company in Richmond, took his family out from the small town of Coles Point, Virginia—about 15 miles downriver from Muse—on a boating excursion. Green's son and his son's friend were water-skiing at about 1:30 p.m., when Green's wife looked up and saw the monster. The boys "came scrambling" out of the water, and Green took his boat closer to the creature to get a better look, but it evaded the boating party. The result of the encounter was a description of a creature that hewed more closely to the "classic" monster profile of 1978 than Muse's encounter had done—complete with humps and vertical undulations. Green's object even left a wake and the occasional whitecap.[29]

Where the Muse and Green sightings contrasted most sharply was in the backgrounds of the people coming forward to the papers. Green and his party were unabashed tourists, and his commentary on the encounter neatly counterpointed that of the indigenous Muse. "That's sort of my second home down there," Green told Cavedo, "and I'm real curious as to what's out there." But while readers in the Northern Neck might have shared Green's curiosity, his legitimacy was a harder sell. When Green told the papers that he had "spent a lot of time on boats down there" and was "used to seeing a lot of strange sights, but never anything like this," the words lacked the resonances of deep experience that came with Muse's plainspoken, rustic wisdom.[30] It was easy to dismiss talk of sea serpents in the Potomac when it came from frivolous tourists

Illustration of the monster accompanying a roundup of the Potomac River sightings, August 10, 1980, by Martin Rhodes, staff. *RICHMOND TIMES-DISPATCH*

and outsiders; contradicting a four-square pillar of the community like Muse was a different kettle of fish entirely.

"Bless Goodwin Muse," wrote Charles McDowell in a sardonic column for the *Times-Dispatch* on July 1, 1980. The farmer had resurrected the Potomac monster just in time to distract Americans from the stuffiness of the 1980 presidential election. "We needed this sea monster desperately," McDowell quipped. "Now there is hope, even as we prepare for the Republican and Democratic national conventions in July and August." The reserved Muse's chaste and matter-of-fact sighting report exemplified a kind of dignity that answered the overblown politics inside the Beltway. "The circumstances of the first sightings of sea monsters are very important," wrote McDowell:

> The sighter has to have credibility if the sea monster is to amount to anything. Goodwin Muse is ideal. His name alone is a gift to all of us in the summer of 1980. Even if you don't notice the name the first time, there is subliminal strength in Goodwin Muse—quiet, thoughtful,

> benevolent authority. At the top of his other qualifications, he is a farmer, an indigenous person, a resident of Westmoreland County for 59 years. Consider how much better this is than if he were some excitable vacationer from Richmond, or a powerboat jockey from Hampton, or, perish the thought, a traveler on a motorcycle from Colonial Beach. Goodwin Muse not only lives in Westmoreland County but his farm is at the George Washington's Birthplace National Monument, for heaven's sake!

McDowell expressed his delight and surprise that Muse's sighting was corroborated by four fellow witnesses, "and the rest of us," he gushed, "can find inspiration in their unity and perception." Contrast Muse and his wholesome companions with a woman who also claimed to see something in the Potomac on the same day as the Muse sighting, McDowell suggested. "She was a Richmond resident who has a cottage at Coles Point," he wrote disapprovingly. "I am prepared to believe this person saw an otter," McDowell joked,

> but this would not influence me to believe that Goodwin Muse saw an otter. It seems to me quite possible, as a matter of fact, that the Richmond resident saw a sea monster and thought it was an otter. Since we are skeptical of summer people—as opposed to indigenous persons—who claim to have seen sea monsters, why shouldn't we be skeptical when they report otters? I will stick with Goodwin Muse, his binoculars and his friends.[31]

Believing Is Seeing

Like Lyons before him, McDowell wore his Menckenian skepticism of the provinces a little too much on his sleeve. But contained within his observations were some home truths about the whole Chessie phenomenon. Whatever McDowell might have believed about Muse's rural plainspokenness, the farmer's position on Chessie was pretty abnormal, and outsiders really *were* the ones who usually spotted sea monsters. Muse aside, Chessie sighters had been almost uniformly nonnatives: tourists just passing through, or newcomers settling into new homes in waterfront developments. Their dogged reliance on sight alone as the basis for proving the monster was real marked each and every one of them as a new arrival in the area—outsiders who did not share the existing culture and were almost as alien to the Northern Neck as Chessie itself. Although the newspapers liked to play up the sightings as something that could only happen in the backward byways of darkest rural Virginia, there was no way to ignore the very clear divide that existed between suburban eyewitnesses and native naysayers, nor the growing tension between the two groups as their worldviews increasingly clashed with each other.

The fault lines were evident as early as the first appearance of the monster. At the time he made his sighting, Donald Kyker and his wife, Ann, had only recently moved to their waterfront home in the Bay Quarter Shores retirement community. Donald Kyker told Anne Hazard in 1979, "It's a little peculiar in the community. We're new here. We have the feeling that some of the local people don't like a story centered here that could be false."[32] He was probably on to something. In June 1980, Carolyn Haynie, a marina worker and lifelong resident of Coles Point, Virginia, could barely contain her contempt when she told Cavedo, "People come down from the city and don't know what they're seeing. You could tell them it was anything and they'd believe it." Haynie's sympathies clearly lay with the native watermen she dealt with on a daily basis, none of whom, she claimed, had ever laid eyes on the monster. "I certainly would think they would see it if something was out there," Haynie told Cavedo. "As many nets up and down this river, I would think that sooner or later it would get hung up in something. You've just got to look at the logical points against it." As far as she and other locals were concerned, Chessie was not even worth talking about. "Nobody here ever says anything about it," she claimed, "unless city people come in and get all up in the air about it."[33] As Cavedo concluded in an August 1980 article, "To most Northern Neck natives living along the Potomac River, Chessie never existed. Just a city slicker's high-strung imagination."[34]

Haynie's faith in her local watermen was a natural response among natives presented with Chessie sightings, and it reveals another facet of the conflict between natives and newcomers: people who worked on the water versus people who played there. Chessie was a symptom of a larger transformation sweeping across rural America in the period after World War II, in which white-collar consumerism displaced existing, working-class economies. Around the ring of the Chesapeake Bay, this process played out in the proliferation of waterfront tourism, recreation, and residential development at the expense of the traditional seafood and farming industries. With tourism becoming a dominant industry along the waterfront of the region, and suburban development encroaching on once-isolated communities, places like the Northern Neck of Virginia gradually saw the transformation of their entire culture and lifestyle.

Because they were nearly all newcomers to the area, the odds were that Chessie sighters were in the Northern Neck to recreate in the water, rather than to work on and in it. The act of seeing Chessie, then, became one by which newcomers imposed *recreation* on the water as its main purpose. Recreation need not mean only playing—fishing, boating, skiing, and so forth—although eyewitnesses were very often engaged in those activities when they had their encounters. Many sighters were simply enjoying the view when Chessie slithered past them. Their suburban ability to appreciate the water gave newcomers the idea that they could pronounce meaningfully on its contents, whatever local watermen might say. Donald Kyker, for instance, felt that his World War II and CIA observational training made him especially well qualified to identify what others could not. When confronted by skeptics, he would demand of them, "How long do you spend actually looking at the river?" Wayne Lawson, a car dealer from Tappahannock, Virginia, who self-identified as a monster hunter, told Hazard, "Crabbers and fishermen have not seen it because 'they're not sitting there, scanning the river. They're looking at what they're doing. Plus,' he said, 'they've got their engines running. If they were to go out there and shut their engines off and look, they might see it.'"[35]

Lawson's suggestion that watermen ought to stop working so they could contemplate the water vividly demonstrates the way that the priorities and mindset of tourism and suburbanization crowded out industry as the primary method for experiencing the water. It also shows how important visuality and aesthetic appreciation were to that shift. Those themes were especially apparent in a Chessie sighting reported by a Richmond resident, Mrs. J. E. Hennaman, in March 1980. Hennaman was headed to dinner at the Harbor Light Restaurant in Hopewell—a Richmond suburb about 60 miles southwest of the Northern Neck—when a snakelike object in the nearby Appomattox River caught her eye. A self-professed "ardent fisherwoman" who had "never seen anything to equal this," Hennaman ran into the restaurant and shouted, "Look outside, the Loch Ness Monster is going down the river." Alas, she was not able to arouse much interest among her fellow patrons. "No one would get up and look," she grumbled when interviewed by the *Times-Dispatch* several months later. Nonetheless, Hennaman believed wholeheartedly in the creature's existence—*and* her ability to discern between it and typical river phenomena. "Some watermen say the water plays tricks on your eyes," she told the paper. "I know what I saw, and the water wasn't playing tricks on my eyes."[36]

If merely seeing something that looked like a sea serpent was now the only standard by which Chessie could be verified, then natives of the Northern Neck were in for a long ride. Even the press recognized the futility of relying on the general public in this way, and in the waning days of summer 1980 newspapers practically pleaded with Northern Neck residents to photograph the monster. "The thing is reported to have been seen by dozens of riverfront residents in Westmoreland and Northumberland counties," the *Times-Dispatch* complained, "but no one—not yet, anyway—has come up with a glossy print of proof." Even Virginia governor John N. Dalton advised readers succinctly, "Take a picture."[37] On July 8, a local paper, the *Westmoreland News*, offered a $50 reward for a photo. Later in the month, the reward was doubled.[38] But no images ever surfaced, and the 1980 sea monster season eventually ended with a whimper.

Ironically, better, if not conclusive, evidence of the creature would not take long to come to light. But that evidence would come not from the Northern Neck, nor indeed any part of the Potomac, but instead from many miles up the Chesapeake Bay, near Kent Island, Maryland. Chessie's change in venue would signal a sea change, not only in sighting locations, which would take place predominantly in the upper bay from now on, but also in the meaning attached to the creature going forward. To be sure, Chessie never lost any of its connections to tourism, consumerism, and suburbanization, but the creature also *gained* some extra dimensions when it swam into these new waters. As it would turn out, 1980s Maryland was an ideal habitat for a sea monster in need of a good home.

Chapter 2

A Change of Scene

THE GOODWIN MUSE ENCOUNTER notwithstanding, the summer of 1980 turned out to be less fruitful for sea monster sightings than 1978 had been. "What has happened to Chessie?" a disappointed *Richmond Times-Dispatch* writer asked in mid-August. The monster seemed to have vanished suddenly and completely from the Northern Neck.[1]

Chessie had, in fact, made a change of address, although it was not readily apparent at the time. Even as Virginians questioned whether monster fever had finally subsided, the creature itself was preparing to make a splash far away to the north. With the whole of the Chesapeake Bay to explore, Chessie's departure from the Old Dominion state was probably inevitable and would quickly become more or less permanent. The move also signaled big changes in the monster's narrative. Advocates for Chessie's existence—the "Chessie faithful," as it were—emerged to refine and expand the creature's mythology, tracking its encounters with the public and developing theories to explain its seemingly random appearances and, maybe more pressing, its *disappearances*. They insisted that scientists and fisheries experts take sightings as seriously as they themselves did.

Although they were not often successful at influencing officials' view of the creature, the "faithful" did, nonetheless, have a hand in remolding Chessie for audiences in Maryland and the wider Chesapeake. Their efforts ensured that the sightings in the Potomac and subsequent sightings in the bay were connected—that is to say, that the Chessie of Virginia and the Chessie of Maryland were considered to be of a piece. In truth, however, eyewitness descriptions from the Potomac often diverged significantly among themselves, let alone from the later Maryland encounters, which in some cases really do appear to have involved a different animal entirely, if an animal at all. Continuing Chessie's story in Maryland waters required a peculiar persistence and a will to believe, both of which the creature found in abundance upon crossing the state line. Retooled for a new decade and a different audience, by people with different sensibilities, this new and improved Chessie would grow into a creature of great staying power.

A Rampant Yul Brynner

No doubt Chessie was the last thing on the minds of Trudy and Coleman Guthrie as they returned home from a leisurely sailing trip aboard their sloop *Impasse* on September 13, 1980. Little did they know, however, as they cruised out of the mouth of the Miles River and into Eastern Bay on this pleasant Saturday afternoon, that they were on a collision course with the monster and a brief moment of local fame. Trudy was at the wheel when she spied what she initially thought was a crab pot bobbing on the water's surface about 10 degrees off the port bow. But the

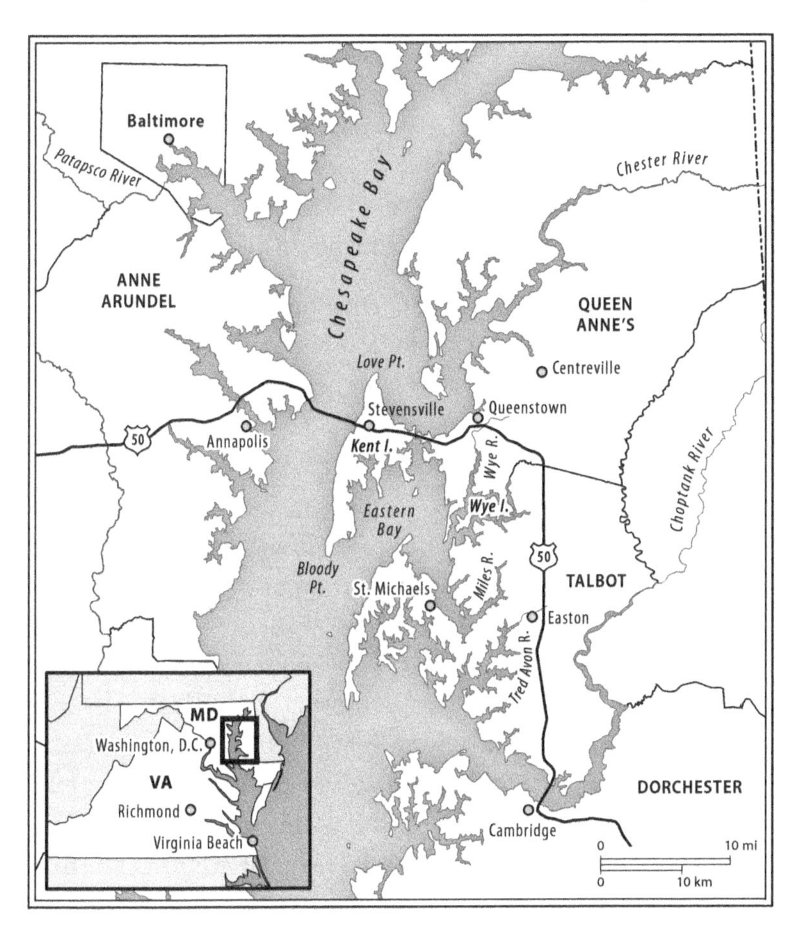

Maryland and the Upper Chesapeake Bay.
ERIN GREB CARTOGRAPHY

object was quickly revealed to be something far less mundane, as the *Times-Dispatch* reported:

> At first, it looked like a human head and shoulders. She then thought it was a skin diver in a tan wet suit, and her husband thought it was a body. The creature continued to rise, however, exposing a large trunk. On its head was a round object that Mrs. Guthrie said could have been an

> eye, blowhole or orifice. Its head was larger than a human's, juglike, and looked as if it were covered with a stocking and faced head down and horizontally, she said. Also, there was no noselike object visible to the Guthries. Between the head and torso was a 14-inch neck, appearing from the side to be 12 inches in diameter.

The Guthries' encounter lasted about 15 seconds, and they got a *clear* look at the creature for less time than that. But Trudy felt certain it was not a seal or stingray, both of which she was familiar with, and neither of which matched what she had seen. "It could have been many things," she told the paper. "Whatever I saw was not indigenous to the Chesapeake Bay." It is worth pointing out that, with her description of a thick trunk boasting a bulbous head that "looked like a rampant Yul Brynner," tan skin, and movement through the water totally unlike that of a snake, what Guthrie claimed to have encountered bore few of the hallmarks of earlier Chessie sightings, nor did it resemble native bay fauna. Indeed, with hindsight, the Guthrie creature seems so radically different from what was seen in the Potomac that it almost certainly had to be a different animal entirely. Nevertheless, the press drew the connection between the Guthrie sighting and its predecessors, enthusiastically promoting the link in spite of the divergent descriptions.[2]

The pivotal element in the encounter was Trudy Guthrie herself. Of all the people in the Chesapeake Bay whom Chessie could have shown itself to, it happened to pick the daughter of Reginald Truitt, founder of the Chesapeake Biological Laboratory at Solomons Island, Maryland. If Goodwin Muse had won the silver medal for witness credibility purely on the basis of his wholesome rural charm, Guthrie's family connection handily nabbed her the gold. As a Maryland fisheries official enthused, "Trudy Guthrie's father is probably one of the foremost marine biologists alive today. She's far better trained than the average layman."[3] It also helped that Guthrie was an experienced sailor and a freelance photojournalist who had contributed to several Maryland newspapers. She was also an Eastern Shore native and the granddaughter of former Maryland governor Emerson Harrington, qualities that paralleled Muse's cachet in the Northern Neck—but surprisingly, the papers never brought up these facts. Perhaps they were deemed unnecessary, given Guthrie's impressive bona fides when it came to marine phenomena.

Ironically, despite her background in photojournalism, Guthrie was without her camera when the creature, or whatever it was, broke the water's surface—a circumstance that brought her great consternation at the time and when she recalled the incident more than 40 years later. It was, she recalls, "one of those things you can't believe. I had been filming; I'd been snapping shots of the bay, of the weather, of everything from sunsets to foliage, of sailboats, of skipjacks, of people oystering, of beautiful scenes in the mist, and here I have the opportunity of a lifetime almost in my lap, and my camera's below."[4] In lieu of a photograph, Guthrie's scientific and sailing instincts kicked in. She immediately pulled out her ship's log and drew what she had seen, depicting it in stages from the first appearance of its

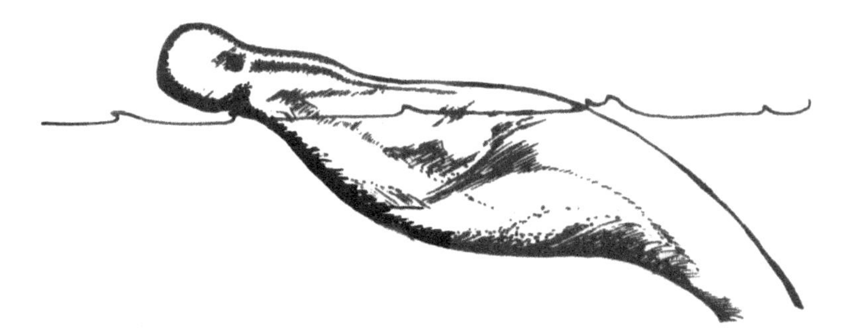

Trudy Guthrie's sketch of the creature she saw in Eastern Bay in September 1980.
COURTESY OF TRUDY GUTHRIE

"head" until it dove out of sight, with notations describing the object's motion, location, and other pertinent nautical information. Another sketch showed what Guthrie imagined the "complete" creature might have looked like.

Despite the press's excitement at her family connection, Guthrie's father encouraged her to keep mum about what she had seen. "His first comment to me was 'Don't tell anybody, because they'll immediately assume it's Chessie. Chessie's story is going wild right now.'" Initially, she and Coleman agreed not to talk about the encounter, but Guthrie's scientific curiosity ultimately caused her to come forward. There were 200 other boats out on the bay the day she made her sighting. Had any of them seen something similar? "I simply *had* to know if anyone else saw it," she remembers.[5]

As Guthrie's father had predicted, the papers seized on her encounter as another episode in the developing story of Chessie. Her background in journalism, her respected scientist father, even the care and precision with which she recorded and reported her sighting—all became threads in the fabric being woven out of the sightings. The *Times-Dispatch*, for instance, took great pains to demonstrate the meticulous way Guthrie measured and recorded all of her observations on the day of the encounter, presenting her as intellectually cautious and reluctant to engage in hyperbole or sensationalism. "Don't bill this thing as a monster," she stressed to the paper. "It's very explainable, but not indigenous to the Chesapeake."[6]

Guthrie's credentials were even more pivotal for Maryland papers. Breaking the story in the Baltimore *Evening Sun*, outdoors editor and family friend Bill Burton treated Guthrie's detailed and technical report as a clincher for the existence of some kind of bay creature. The Guthries were "experienced sailors," Burton wrote; the description implied not only accurate reporting and steadiness of judgment but also trustworthiness. The Easton *Star-Democrat*, the leading newspaper of the central East-

Illustration accompanying the *Richmond Times-Dispatch*'s story on the Guthrie sighting, September 21, 1980, by Martin Rhodes, staff. *RICHMOND TIMES-DISPATCH*

ern Shore, echoed this seriousness. Part of its article focused on the grim possibility that the Guthries had indeed seen a cadaver floating in the water.[7] Whatever they had seen, it was no laughing matter, but the national press was not so restrained. "It's been a bumper year in the Chesapeake Bay—and not just for crabs," an AP article published in the Annapolis *Capital* wryly noted. "Monsters, sea lizards, unidentified swimming objects—whatever they may be—have been seen in goodly numbers this summer of 1980."[8]

Probably to her chagrin, Trudy Guthrie's encounter proved vital to the extension of the Chessie phenomenon. Her biggest contribution to the emerging legend was to solidify, with Burton's help, the conceit that Chessie was a "friendly" monster.

The Guthries, Burton wrote in his *Evening Sun* account of their encounter, "were neither threatened nor frightened" by what they had seen. As Trudy Guthrie put it to a *Times-Dispatch* reporter, "When we saw it, there was no fear." The paper amplified her assertion, claiming that "there is general agreement that Chessie is friendly, but somewhat shy."[9] In truth, however, no such consensus existed. Back in 1978, the Potomac monster had been presented as, well, *monstrous*. Eyewitnesses and public officials alike regularly compared the creature to the shark in *Jaws* and openly wondered if the creature was dangerous. Only one witness at the time, Ann Kyker, tentatively suggested the creature might not be "scary," and one fisheries official suggested classifying the creature as "friendly" in an effort to protect it from violence. But such reactions were rare compared with the chorus of primal fears the early sightings generated.[10]

Nor had the creature's reputation improved much in the intervening couple of years. The idea that by 1980 Chessie—an as-yet-unidentified aquatic creature that by most accounts seemed to resemble a giant snake—was not only benign but now somehow *friendly* was a figment of the media's imagination. "In none of the incidents reported have the creatures seemed the least bit aggressive, or even concerned in their encounters with humans," Burton wrote in October 1980.[11] The proposition seemed plausible only because Trudy Guthrie's forceful and dispassionate assessment of the creature made it seem misunderstood and unthreatening, a curiosity to be studied rather than a danger to be eliminated.

In a wider sense, however, the Guthrie sighting inspired a whole bay monster culture that eventually became attached to Chessie. It caused Burton to begin soliciting sighting reports from readers around the bay and to scour the *Sunpapers* archive for historical sea serpent accounts to go with them. "There is a long history of unusual 'monsters' of the Chesapeake Bay region; some not recorded undoubtedly go back to the days of the early

Indians," Burton concluded, rather imaginatively, in a November 1980 spread on the Guthrie sighting for *Chesapeake Bay Magazine*.[12] However fanciful his assertion, it points to the way Burton imagined the phenomenon he was tracking. Whatever variations existed in reports—and the Guthrie sighting seemed to suggest they were myriad—all of the "monsters" seemed to be related, if not the same, and belonged to the same lore.

In the months following the Guthrie sighting, Burton gradually deepened the urgency and mystery of the monster, using local reports that came to him, however dubious or extraneous, as the catgut to suture Guthrie's creature together with the earlier sightings in the Potomac. Some of those reports boasted the same level of detail as Guthrie's account, and some strongly blurred the line between verifiable journalism and hearsay. Witnesses' credibility was always key, and that credibility, in turn, indexed to knowledge of the water—especially in the instances where actual details were thin on the ground.

In October 1980, for example, Burton mashed up the Guthrie sighting with a report that came in from Steve McKerrow, a colleague who was "an experienced boater and boating columnist for the *Evening Sun*," and a fish tale from two anonymous "old-time crabbers" whose story came in from a local sports shop owner. The crabbers, described as "knowledgeable watermen" who were "quite sincere," related that they had pulled up one of their crab pots only to find "perched atop it a big and obviously alive being of grayish shade—and with no hair"—that apparently grunted at them. The men believed this creature was an octopus, but Burton lumped it in with all the other accounts.[13]

In a similar column in November, Burton reported on a sighting made in a marina off the Gunpowder River in Joppatowne, Maryland. The story had a familiar refrain: the witness, Linda Worthington, had only just moved to her waterfront condominium when she made the sighting, and, on the basis that "she has been around the water a lot," she insisted that what she saw was not a typical bay animal. What made the account unusual, however, was the way Burton handled it. In his words, Worthington described the creature she saw as "serpent-like" and "somewhat akin" to the Guthrie creature.[14] The contradiction in terms here is glaring: Guthrie's creature was explicitly *not* serpentlike. Surely Burton recognized as much. That he glossed over such a significant detail is surprising and a bit hard to explain, but it demonstrates his central role in winnowing out inconsistencies among sighting descriptions to produce what eventually became the "classic" Chessie mythos.

By the end of 1980, Burton had clearly come to regard the Guthrie sighting, if not the particulars of the monster itself, as the gold standard against which all other sightings should be compared. Much as the monster had begun to gravitate to points north of the southern reaches of the bay, the Guthrie creature had moved from the periphery to the center of Burton's Chesapeake Bay monster lore. It was pretty obvious why: with witnesses of Guthrie's caliber coming forward, there *had* to be something to the sightings.

Writing in January 1981, Burton identified another eyewitness on par with Guthrie—indeed, a witness who was so reliable that he seemed to see *the same monster*. Virginia Beach businessman Bob White had been sailing southward from Windmill Point toward the Chesapeake Bay Bridge Tunnel on September 7, 1980—a week before the Guthrie encounter in Eastern Bay—when he saw something in the water that was identical "almost to the letter" to the creature Guthrie described and sketched. "It matched perfectly what I saw," he told Burton. "Two holes about the size of 200 mm shells on the side of the head, no eyelids or eyes as such; greenish brown, no fins, scales, or mouth evident, no nothing, but definitely a living thing."

In its style and much of its content, the article was a conscious redux of Burton's report on the Guthrie encounter. As with Guthrie, Burton paid special attention to White's credentials. "Write him off as a landlubber," he wrote, "[and] he won't hesitate to inform you he's been a boatman since he was 16, has cruised the Chesapeake 14 to 15 years, was for two years commodore of his Coast Guard Auxiliary Flotilla, last year was its public information officer, and this year is educational officer." Accordingly, White's sighting was documented with details that almost certainly came straight from his ship's log.[15]

And just in case *one* eyewitness to a creature like Guthrie's was not enough, Burton produced *another*, a friend of White's named Harry Fremd, who claimed to have seen a creature of the same description too. "No untrained eyes on Harry, vice commodore of the auxiliary, retired Navy fighter pilot whose military fortunes as a flight commander depended on keen observation," wrote Burton in a statement reminiscent of reports about Donald Kyker back in 1978. The available evidence, backed by some very convincing witnesses, seemed decisive. "Still think all of those sea creature stories are figments of imagination?" Burton asked readers.[16]

The Chessie System

Whatever convictions Burton might have held, they were not necessarily shared by anyone outside the circles of the Chessie faithful. For all that the Guthrie encounter was well documented and seemed credible, there was still no certainty about what Guthrie had seen. Clearly, Burton thought there was an exotic creature of some kind inhabiting the bay, but even if there were, did that mean the creature was a cryptid? What if Chessie was a nonindigenous animal just passing through?

Burton's beliefs notwithstanding, the Guthrie sketch—for a while the only available image of the monster—seemed to point in that direction. Looking at the sketch without any preconceived ideas about sea serpents or other cryptids, it is hard to deny the resemblance to a manatee. Guthrie herself believed that the endangered "sea cow," an animal native to Florida and other tropical regions, was a strong contender for what she had seen that September day. But manatees were not known to migrate as far north as the Chesapeake, and Guthrie felt that what she saw was too large and long to be a manatee.

Another possibility emerged a few days after the Guthries' story appeared in the *Star-Democrat*. Charles Kepner, of St. Michaels, Maryland, wrote a letter to the editor claiming that he and his wife, Susan, had seen the same creature on August 15 (a month before the Guthrie sighting), while fishing from their sailboat. Kepner contended that what they had both encountered was a harbor seal that had followed a school of bluefish out of the Atlantic and up the bay. Kepner's letter set off a brief debate in the paper over which animal—manatee or harbor seal—was more likely to be Guthrie's creature. The verdict, perhaps surprisingly, seemed to be in favor of the harbor seal. The Atlantic coast and the Chesapeake were too harsh an environment for manatees, explained W. P. Jensen, Maryland's director of tidal fisheries, whereas a harbor seal would find the brackish bay water inviting.[17] Trudy Guthrie was quick to respond: "She is quite familiar with seals—and the creature she saw did not appear to be one, nor did it swim like one," wrote Burton after an interview with her.[18]

The harbor seal debate signaled a bit of backsliding regarding Chessie. If the Guthrie monster was really not a monster, then the lore Burton had been so painstakingly building would unravel completely. Maybe Maryland and the upper bay really had nothing to do with the Potomac River sightings after all. Soon the point was no longer academic. On September 28, a report came in from Smith Point, near the mouth of the Potomac and some 80 miles south of the site of the Guthrie encounter. On this occasion, about 25 people in four charter fishing boats saw the creature, and this time Chessie appeared more like its old self, presenting in the form of a "serpentlike thing." There was a slight variation in the description, though: some witnesses claimed its dorsal region was "spiked." This trait strongly suggests witnesses saw a sturgeon, a bony-backed and vaguely prehistoric-appearing fish plentiful in the lower Chesapeake, but the resemblance was close enough to the "classic" Chessie to get by.[19]

In any event, Chessie's newly barbed back was a small price to pay if it meant restoring the monster (mostly) to its pre-Guthrie description and Virginian habitat. Certainly the *Times-Dispatch* was in more familiar territory when it reported on the sighting: the captain of one of the boats, Donald Markwith, obligingly told the paper that "he was skeptical of reports that a strange creature was roaming the Chesapeake and its estuaries, until Westmoreland County farmer Goodwin Muse said he saw it." In Virginia in 1980, no Chessie story was complete without returning to the Muse touchstone. And in this Chessie encounter, there was *twice* the Muse connection: in one of the many uncanny coincidences in the Chessie saga, one of the other boat captains who saw the monster at Smith Point, Bill Jenkins, just happened to be Muse's brother-in-law.[20] Had Virginia won back ownership of the monster in its brief tug-of-war with Maryland?

Alas, the results were inconclusive: after the Guthrie and Smith Point episodes, Chessie abruptly vanished again, leaving the monster debate in a state of flux and uncertainty. In the absence of new sightings, and tantalized by Guthrie's sketch, the media obsessed over the lack of a clear image of the monster. During the summer of 1980, local newspapers in the Northern Neck had offered a reward for a photograph, and in nearly every one of its news articles, the *Times-Dispatch* quizzed eyewitnesses about whether they were carrying a camera at the time of their sightings and if they would do so in the future. Many sighters insisted they would keep a camera on hand going forward, but a persistent *lack* of cameras among sighters became one of the more amusing aspects of Chessie lore. "So far, Chessie has eluded photographers," reporter Albert Oetgen complained in November 1980. "Several people who normally carry cameras on their boats reported that, for one reason or another, they didn't have them when they saw the creature." As we have seen, Guthrie was one of those people. "I wouldn't have had time to take the picture anyway," she admitted at the time. "I couldn't get my mouth closed."[21]

Although the papers in Virginia might have thought they were on the cusp of a real breakthrough when it came to visual evidence, and in spite of the monster's brief spell of recidivism in late September, Chessie really was in the process of changing its address. By widening the frame of its adventures from the environs of the Northern Neck to the complete expanse of the Chesapeake Bay and its tributaries, the larger portion of which is located in Maryland, Chessie was effectively leaving the Old Dominion behind. Whatever else it augured, the Guthrie sighting symbolized the monster's immigration into Maryland waters. Reflecting on that day recently, Guthrie recalled that the creature she saw and sketched was clearly "on a course." Tacking their way home that afternoon, the Guthries' course took them south. The creature, she says, "was heading north."[22] A change of venue was definitely in its future—whatever *it* turned out to be.

In the enterprise of relocating Chessie to Maryland, Burton was a willing participant. His reporting on the Guthrie sighting widened the frame on Chessie's travels, locating them in what he called the "Chesapeake Bay complex"—an interconnected network of land and water that transcended political boundaries, with Maryland as its central axis.[23] No longer constrained to the Potomac, the Chessie of Burton's imaginings was more wide-ranging, inhabiting the system of waterways that made up the Chesapeake watershed. If Chessie really were an aquatic animal, Burton's view was probably accurate, but it subtly moved the spotlight off the piece of the "complex" with which the serpent was most connected, and did so in spite of most of the evidence available at the time.

Burton's voice was unusually influential in this process, not least because he brought a sheen of legitimacy to studying Chessie. Widely known in Maryland journalistic, political, and outdoor-sport circles, Burton was a highly respected authority on the Chesapeake. Born in Rhode Island in 1927, he had served as a Seabee in the Pacific in World War II. He became a newspaperman after the war, doing stints in New England, Alaska, and Nebraska before taking the job as outdoors editor for the *Evening Sun* in 1955. In his nearly 40-year career with the paper, Burton became a household name, not only due to his columns and articles on the bay and Maryland's natural resources but also through appearances on television and radio.[24] His interest in Chessie gave the monster—or investigation of the monster, at any rate—a respectability that it had not enjoyed before. His celebrity raised its profile immeasurably.

Despite his reputation, Burton's growing fascination with Chessie still made him (and maybe the monster too) the butt of a few jokes. Responding to Burton's *Chesapeake Bay Magazine* spread in November 1980, *Evening Sun* columnist James H. Bready laughingly depicted Burton as an obsessive cryptozoology nut, desperately soliciting evidence of Chessie and tracking down any lead, no matter how distant or far-fetched. "Bill Burton is convinced," Bready joked. "You should understand, this is a moment": after a career spent hearing fish stories, Burton had decided that Chessie was the real deal, and the ultimate nature writer's scoop. "Even if it happens a hair below the Virginia line," Bready joked, "even if the science writers promptly swarm in—what a story!"[25] The caricature, no doubt intended as a jape between old colleagues at the *Sunpapers*, contained a grain of truth. Burton *was*, in fact, compiling monster sighting reports, and in his magazine article, he had asked readers to send in information about their own encounters. He was also willing to follow up on sighting reports, especially older ones that might establish Chessie's presence in the bay before 1978.

Burton's contribution to the Chessie mythos, however, was no joke, and Bready's throwaway comment about the Virginia line shows just how influential it was. Suddenly, Chessie belonged to Maryland. The assumption threaded all the way through Bready's column, as it did, for instance, when he speculated on why the monster might periodically surface: "The bay beast may or may not admire Maryland's littoral," he wrote, casually ignoring the fact that nearly all Chessie sightings had taken place in Virginia. Bready even took issue (humorously) with Chessie's name, and both his reasoning and his suggestion for an alternate name distanced the monster from its Virginia roots. "Chessie" was unacceptable since it confused the creature with Chessie System, Inc., the holding company that operated several railroads in and around Maryland, including the Baltimore and Ohio Railroad, one of the oldest lines in the country. Boasting one of the most recognizable brands in the region, the Chessie System was deeply connected to Maryland identity, especially in Baltimore, and was as much in the news in the 1970s and 1980s, the era of deregulation, as its serpentine namesake. To avoid misunderstanding, Bready modestly proposed that the monster instead be dubbed "Bessie." Why? Because "it'll sound like Baysie."[26] Once again, in a very subtle way, the monster was distanced from its history in Virginia—but also in a way that made Maryland the fulcrum of life in the Chesapeake.

It seems fair to assume that neither Burton nor Bready was looking to rewrite the Chessie narrative, much less put Maryland at the heart of the Chesapeake, but that was what was happening, and Burton was largely responsible for it. The gravitational force of his prestige was powerful enough to attract the monster out of its path through Virginia and pull it deep into Maryland, where the creature could represent both states, as well as the larger bay system that it now seemed to occupy. The lens on Chessie's story had gone to wide-angle, and, pulling back, what had seemed to be the center of the field of view was really on its periphery. When the coveted visual evidence of Chessie finally appeared, in mid-1982, not only the monster but Maryland would be at the point of focus.

Even with Burton on its side, however, Chessie failed to return to *any* part of the Chesapeake in 1981. In April, the *Times-Dispatch* optimistically tried to drum up sightings with a retrospective article reminding readers that Chessie was a seasonal visitor to the Northern Neck, and that two years of strong sighting reports, 1978 and 1980, had been separated by a year of none at all in 1979.[27] The article was ambivalent about prospects for 1981, and its doubts proved well founded when no reports came in by year's end. Writing in December 1981, a pessimistic Albert Oetgen grumbled that Chessie "apparently decided to spend 1981 in seclusion." It really did seem now as if the monster manifested itself only in even years, a point that at least offered hope for 1982. But what if the even-year theory was just self-delusion? "Is Chessie real?" a crestfallen Oetgen wondered.[28] It was on this minor note that the monster's career in the Northern Neck came to a close.

Something Literally Marvelous

The proposition that Chessie was a biennial visitor was probably more a product of wishful thinking than the scientific method, but it was nonetheless proved spectacularly true when the summer of 1982 rolled around. On June 30, 1982, the *Kent Island Bay Times* broke the news that Chessie hunters had so long awaited: a couple living on the north of the island, at Love Point, had not only seen the monster, they had *videotaped* it. It was a watershed moment that probably rescued Chessie from drifting into obscurity after the recent lack of sightings. Now the monster burst back into the local consciousness, becoming an overnight celebrity across the region and beyond.

The sighting took place on the evening of Memorial Day, May 31, 1982. Robert and Karen Frew had invited some friends over for a small party, and the group was in the Frews' dining room, which looked out on the bay. Charlotte Rosier, the Frews' neighbor, happened to look up and was the first to spot something going by the bulkhead at the end of the yard, about 200 feet away. "Look at that big thing floating there," she exclaimed, pointing at the object. The "thing," it turned out, was not floating, it was swimming—against the current. It dove and surfaced while the Frews and their guests watched, during which time they estimated the creature to be 30 feet long, based on a comparison with the Frews' swimming pool, which was 36 feet long. Rosier's husband, Steve, told the papers the creature left him in shock. "At first I just stood there and didn't move, then we all ran down to the bulkhead to get a better look."[29]

Robert Frew eventually thought to fetch his camcorder and film the creature. From an upstairs window, he was able to record three or four minutes of what the *Bay Times* called "fairly well-focused color shots, complete with sound recordings of reactions by spectators." From his perch, Frew was able to look up toward Love Point, where a group of swimmers seemed to be frolicking right in the creature's path. The situation reminded him of a horror film: "I was afraid to take the camera off it because I was thinking in the back of my mind, 'Any minute now we're going to start seeing these people disappear one by one.'" Although the Frews and Rosiers tried to warn the swimmers to leave the water, the swimmers did not hear them. In the event, the creature ignored them completely, apparently diving underneath them and resurfacing farther away. Although he drove to the tip of Love Point to see if he could spot the monster, Frew said he never saw it again.[30]

Even though the encounter could only have lasted maybe 20 minutes from start to finish, the Frew sighting became *the* defining moment in the Chessie story. As an acerbic *New York Times* editorial observed, "This being the 20th century, we demand higher standards of proof than mere seamen's tales or maps tagged 'Here Be Monsters' in corners the cartographer wasn't sure how to fill in. Our monsters must be recorded on film."[31] Now that film took center stage, causing a stir that brought the monster national and international fame on par with that of A-list cryptids like the Loch Ness Monster, and conferring on the Frews a celebrity that they quickly came to despise. The Frew film also transformed the debate about what Chessie was. There no longer seemed to be any debate as to whether the creature actually existed. The question now was, what *was* it?

It is easy to be swept away by the prospect of real, honest-to-goodness visual evidence of Chessie and forget to take a moment to ask why the monster was filmed in this particular place and time, by these particular people. The questions of time and place are easily resolved: the serpent's long-standing modus operandi was to appear in the warmer months, when people were out on the water, or in it, or looking at it. In these respects, the Frews and Rosiers were following the same old Chessie playbook that had been around since Donald Kyker first saw the monster in 1978. There are farther-reaching questions relating to time and place, however, such as why the sighting happened in 1982 and not earlier or later, and why it happened specifically on Kent Island. We will defer those until later, when more sense can be made of the context in which the sighting took place.

So why this particular group of people, then? The venue had changed from the Northern Neck, but the Frew sighting replayed the familiar story of outsiders moving into a rural area and remaking the water and the waterfront in their own image. Like the Kykers, Smoots, and other eyewitnesses before them, the Frews were relatively new arrivals at Love Point, an area of Kent Island that had been developed in the previous couple of decades from open farmland and shanties into an increasingly exclusive residential area inhabited by affluent people. Although the family had moved to the island from Annapolis in 1977, Robert Frew had been boating around Kent Island for 17 years. It was a point he stressed in news reports as evidence of his awareness of what the water ought and ought not contain and, as we have seen in Virginia, a pretty common trait among Chessie sighters in rural places.[32]

Probably the most revealing element of the Frew encounter, however, is the casual appearance of a video camera, a luxury so rare on the Eastern Shore of Maryland (and probably most rural places) in 1982 that it might as well have been a mythical animal. Consumer video cameras (or camcorders) were just becoming widely available and cheap enough for regular people to own in the early 1980s, but they still required, and showed off, a certain amount of affluence. A mid-Shore electronics dealer advertised one for sale a few months after the Frew movie that would have set buyers back almost $800. For comparison, using 1979 data adjusted for inflation, the median household income in Queen Anne's County, Maryland, where Kent Island is located, was in the neighborhood of $13,500 (in 1982 dollars). For the average Kent Islander to film Chessie that year, he or she would have had to fork out about three weeks of income. For the video to have been made in the Northern Neck, in Northumberland County, where the median income was about $10,600 (in 1982 dollars), the sacrifice would have been even greater.[33] In either locale, the presence of a video camera would have differentiated its wielder from most people because of the disposable wealth it implied. It is perhaps no surprise, then, to discover that Robert Frew was a laser salesman—a profession and industry far removed from what most Kent Islanders did for a living.

In 1982, however, change was on the horizon. As we shall see in chapter 4 when we revisit those lingering questions of time and place, Frew's video came along just as Kent Island was in the throes of an economic and social revolution that was playing out in the politics of waterfront development and causing increasing tension within the community. Affluent newcomers seeking a slice of the good life created a market for residential and commercial development that seemed to crowd out poorer natives, many of whom made their livelihoods as watermen. Even less recent arrivals in older developments, like the Frews, helped to tip the island culturally and economically toward mass suburbanization. The presence of Chessie on the Love Point shoreline, and on Robert Frew's videotape, signaled the irreversibility of these changes and the ambivalent future many native islanders faced.

Of course, an abstraction like the fate of Kent Island was far from anybody's thoughts that Memorial Day evening at Love Point. When confronted by the spectacle of a sea monster swimming by your home, and the possibility that you might have hard evidence of it, the questions on your mind are not about the historical forces that brought it there. In the immediate aftermath of the sighting, there were much more pressing matters at hand, and they had immediate personal consequences. For instance, should the Frews and Rosiers tell their story? Would anyone believe them? How would going public affect their lives? Because of the videotape, Robert and Karen Frew had another decision to make: Should they share it? Ultimately, both families went to the press, but it was the Frews who took the lead, for obvious reasons. Eventually, Robert Frew contacted a news station in Baltimore, WJZ Channel 13, which sent a reporter to investigate. A few weeks later, a segment of the video appeared alongside an interview with the Frews on WJZ's evening news program, and the couple became instant local celebrities.

Yet fate must have been involved *somehow* with the Love Point sighting, because, in a spooky case of life imitating art, the encounter had been presaged in a novel released—incredible, but true—just the week before! In *Sabbatical: A Romance*, the noted postmodernist author John Barth uses the narrative of a sailing trip up the Chesapeake Bay as a metaphor for a married couple's personal journey. Chessie appears close to the end, when the couple, Fenwick Key and Susan Seckler, are faced with making a pivotal life decision. As they sail from the Western Shore to Kent Island, rounding Love Point on their way toward Kent Narrows, brooding all the while about their future together, Susan sees what she initially thinks is a drowning victim floating nearby. Much as he did for other aspects of the book's plot, Barth drew on actual news reports (he directly references the Guthrie sighting in a footnote), so his literary Chessie encounter bears the classic hallmarks of a real-life sighting: the object in the water is sleek and vaguely serpentine, with few other discernible features; it is like nothing Fenwick and Susan have ever seen on the water before. After cycling through the possibilities, the couple must conclude it is Chessie, and Barth announces, with characteristic metatextual wit, that "a legendary sea-monster swims through our story."[34]

A comical sequence follows, in which Susan attempts to photograph the creature—unsuccessfully, as it turns out, first because she forgets to remove the lens cap on her camera, and then because she is slow to advance her film. The resulting photo "contributes to the annals of marine biology a passably clear 35 mm. color exposure of a foam-flecked swirl on the calm surface of Chesapeake Bay, where our monster has just submerged."[35] The couple thinks they spot Chessie again, in the direction of Love Point, but they cannot be sure, and Susan speculates that the CIA is training sea monsters to keep tabs on her husband, who is a retired member of the agency on the outs with his former employers. In an accompanying footnote, Barth reminds the reader that Donald Kyker, Chessie's first eyewitness in 1978, was also a CIA retiree.

The encounter inspires Fenwick to have an epiphany. Initially, he wants the monster to reappear so that contemplation of its wondrousness will delay the decision he and his wife have to make about their future. "He feels sharply," the novel tells us, "that beyond this narrative diversion, this ontological warp in our story—a fabulous sea-monster in our real Chesapeake!—lies something hard and hurtful."[36] Later, Fenwick realizes that fantastic Chessie is the *answer* to his existential problem, not a bromide or impediment. "Something happened just now, starting with Love Point and our sea-monster," he excitedly tells his wife after his revelation, as he reaches his decision and thinks through what it will mean for the couple's future together.[37]

Why include Chessie? Topicality, principally, although Barth was not to know just how up to date the monster's inclusion would be when the book hit the shelves. A native of Cambridge, Maryland, and a professor at Johns Hopkins University from 1973 to 1995, Barth positively revels in references to local geography, culture, and history in *Sabbatical*, grounding (immersing?) its plot in contemporary events like the CIA tell-all scandals of the late 1970s, the bay's environmental travails, and, of course, the Chessie sensation. Reviewers were generally lukewarm toward the novel, and if they seized on the presence of the monster at all, it was usually in order to chide Barth, as *Kirkus* did, for his "cutesy" excesses. The *Christian Science Monitor* was even harsher, turning the sea serpent metaphor back on Barth and using its indeterminate, chimerical nature as a symbol of the book's inability to be one thing or another. "And it's gone aground in Chesapeake Bay. Alas," reviewer Victor Howes dryly concluded.[38]

In a 1985 interview with George Plimpton, Barth—who was then presumably working on *Sabbatical*'s follow-up, *The Tidewater Tales* (1987), another sailing yarn teasing occasional cameos from Chessie—commented on the serpent's manifestation at the earlier novel's climax. "Early in the planning of *Sabbatical*," he explained, "I knew that in the next-to-last chapter, when the characters sail their boat around a certain point, something extraordinary had to happen, something literally marvelous, but I had no idea what that ought to be until they actually did turn that corner. By then the metaphor was clear enough so that the sea monster, which had to surface at that moment, surfaced in my imagination just when it surfaced in the novel."[39] That the Frew sighting played out in the same place only a week after *Sabbatical* was published is a collision between fact and fiction so postmodern that Barth himself could have written it. If one were looking for "literally marvelous," this strange coincidence was surely it—and it would not be the last one in Chessie's story.

A Riddle Wrapped in an Enigma

As the Chessie faithful had predicted for several years, the Frew recording—actual visual evidence of the creature—was a turning point in the Chessie phenomenon. Not only did it raise awareness of the ongoing mystery beyond the local, propelling the monster phenomenon to national and international stardom, but the video also secured Chessie's status as Maryland's monster. In the wake of that fateful Memorial Day, the creature's amorphous prehistory in Virginia would be increasingly glossed over and forgotten. Few further sightings took place in Virginian waters after 1982, and the ones that did failed to cause much stir. The Frew video also initiated a second phase in the life of Chessie, in which the creature increasingly symbolized the uniqueness of the Chesapeake region and, in turn, the bay's intrinsic connection to the history, culture, and lifeblood of Maryland. Ironically, the first proper visual evidence of the monster brought it in from the hinterlands of myth and sensationalism just in time for it to be transformed into the basis for another kind of mythology.

The existence of putative visual proof of Chessie raised a host of new questions. Was the video a fake? How could that be determined? If it were genuine, what did that mean, and how could the creature supposedly in it be identified? Who would undertake a study of the film anyway, and what authority, scientific or otherwise, would they have to give the public meaningful answers to all these questions? Maryland conservation officials were not exactly volunteering to study—or even acknowledge—the video, and there was no authority within the state charged with the responsibility to investigate strange happenings. The scenario was straight out of science fiction, but reality was unequipped to respond to it.

Enter the Enigma Project, a Reisterstown, Maryland, group established to chase down accounts of strange phenomena in the mid-Atlantic, probe them, and hopefully find rational explanations for them. Friends Michael Frizzell and Robert Lazzara, both engineers at Baltimore-area manufacturing firms, founded the group in 1977, along with several others who shared their interest. "We all had scientific and technical backgrounds," Frizzell remembered in 2005, "and we could see that when certain controversial stories—fringe, scientifically fringe stories—hit the media, . . . generally these subjects were pooh-poohed, and not really given *any* degree of serious scientific attention."[40] The Enigma Project's purpose was to give those stories the rigorous analysis the group thought they deserved, if not for the public's benefit then to satisfy members' own curiosity.

If the Enigma Project's raison d'être seems familiar, it is because the group shared its roots with much of the same culture that launched Chessie in the first place. One strong inspiration for the group, for instance, was *In Search Of . . .* (1977–82), which presented half-hour documentaries on a wide range of mysterious phenomena, from UFOs, to the Loch Ness Monster, to the disappearance of Amelia Earhart. Another influence was Jim Brandon's *Weird America: A Guide to Places of Mystery in the United States* (1978), a travel guide to strange phenomena in America. "The whole decade of the 1970s were, in my opinion, very active in terms of all kinds of fringe phenomena," Frizzell observed in 2005. "All kinds of really odd newspaper articles, stories, whatnot . . . that occurred concurrently with an increase in all kinds of unusual animal sightings, primarily Bigfoot. . . . It just seemed like a weird decade for strange reports."[41]

At this point, readers might imagine that the Enigma Project was a shoo-in for analyzing the Frew tape—and it probably was, except that under normal circumstances there was no easy mechanism for uniting the two parties. As fortune would have it, however, and in possibly the strangest coincidence in the whole Chessie saga, Robert Frew *already knew* Frizzell and Lazzara. Back when the Enigma Project was first organized in the late 1970s, its members decided to tackle an investigation of mysterious lights reported around Maryland. This enterprise eventually took the group quite far afield, to Brown Mountain, North Carolina, where they intended to examine a famous case of strange lights reported in that area. In order to perform experiments, however, they needed a laser. After some hunting, Frizzell located a laser salesman on Kent Island who was willing to lend them the expensive device. The salesman was, of course, Robert Frew. When news of the video broke in June 1982 along with an interview with Frew, Frizzell could scarcely believe his eyes. "I thought to myself, 'My God! This is the same guy I got the laser from!'" he later recalled.[42]

But for this preexisting relationship, the Enigma Project might never have turned its attention to Chessie at all. Frizzell and Lazzara were peripherally aware of the sea serpent story as far back as the first sightings, but awareness never became interest. "Quite frankly," Frizzell confessed in a 2001 essay, "I must admit that on first hearing of 'Nessie Junior' in 1978, I was skeptical." Chessie, he thought, was nothing more than "the product of someone's overactive imagination . . . now being turned into a poor man's Loch Ness Monster by the media."[43] Whatever his reservations, however, the Enigma Project and Chessie seemed destined to meet.

Lazzara and Frizzell made arrangements to visit the Frews in mid-June 1982, and due to their astounding inside connection, they were received at Love Point "like old acquaintances."[44] The researchers interviewed both the Frew and Rosier families and found their accounts of the sighting to be consistent across the board. Frizzell later wrote that they all impressed with their "credibility" and "the solidity of their evidence," describing the four adult witnesses as "well educated, sincere professionals."[45] The slight class prejudice contained in his assessment should not shock us too much. Chessie's appearances had been exacerbating those sorts of social tensions since 1978—and what other way was there to determine the reliability of a report of this kind but to ask why people might risk their reputations by speaking out? At the same time, the Enigma Project's reputation was at stake too. Examining the Frew film was a big break for the organization, and a high-profile one at that. Pursuing an investigation that was obviously the result of observer ignorance or, worse, a hoax could ruin the group's reputation for quality research. Frizzell and Lazzara needed to believe in the Frews, especially, before they could allow themselves to believe in Chessie.

Belief is a powerful force, though. It seems pretty clear that the Frews *believed* they had seen Chessie, by which they meant a sea serpent of now-classic proportions and description. A suitably serpentine illustration accompanying the *Sunpapers'* story on the Frew sighting reinforced the imagery. "Robert Frew says the 30-foot-long creature he saw in the Chesapeake Bay off Love

Point looked like this," its caption read.[46] But based on newspaper stories, the couple seem to have been predisposed to see Chessie anyway. When the *Kent Island Bay Times* reported on the video, for instance, it noted that Robert Frew "has followed Chessie for many years and has newspaper clippings plastered on his garage wall."[47] Even as he was in the process of filming, he was convinced he was making a movie of Chessie. Although no sound copy of the video is now known to exist, a 1993 news feature recalled him hollering to swimmers in the water near the creature: "Hey! Watch that thing coming at you, guys. It's Chessie!" Karen Frew was apparently less convinced that her Chessie was the "real" Chessie—that is, until Lazzara and Frizzell cracked open an illustrated guide to the world's snake species in front of her. In a case of what one might call self-fulfilling prophecy, she found a match. "I said 'that's it' and the label underneath it said something like an unclassified sea monster," she told the papers.[48]

None of this is meant to imply that the Frews and their guests were not entirely sincere about what they had seen. Nor can we say definitively that a giant sea serpent was not swimming by their house on Memorial Day 1982. The Frews and everybody else involved were faced with two epistemic problems. On the one hand, they had to decide whether the film they took actually depicted anything at all and, if so, if it was alive. On the other hand, they needed to identify that thing, assuming it even existed. It was a riddle puzzling enough to tax anyone's patience,

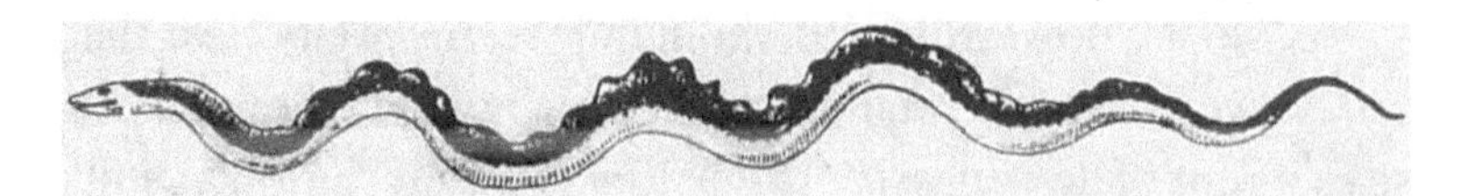

Artist's conception of the creature sighted by the Frew and Rosier families, appearing in the *Baltimore Sun*, July 11, 1982.

so one can hardly blame the Frews for cutting the Gordian knot and assuming it was Chessie. So did everyone else.

The Frews were eager to see the video analyzed by experts in order to identify the creature. As Karen Frew told the *Washington Post*, "I want to find out what the heck it was. I want it classified. I want somebody to put a label on it."[49] Accordingly, Lazzara and Frizzell copied the tape and promised to deliver it to "the most relevant scientific professionals."[50] Like the footage that had accompanied WJZ's news segments, this copy was obtained by pointing a camera at a television monitor displaying the video and recording the output. It was an ungainly process that nonetheless ensured that the original tape—and the video's copyright—remained in Frew's possession. Unfortunately, the resulting image quality was less than optimal even for casual viewing, let alone for close study. Worse, by all accounts, the original tape was eventually lost, so any expert analysis of the Frew film that has *ever* taken place can be traced back at best to an off-monitor copy.

With their copy of the video in hand, Lazzara and Frizzell had one more step to take before they could begin their investigation in earnest: making contact with Bill Burton. As the locally acknowledged expert on Chessie at that time, Burton was the obvious first port of call for determining what the Frew tape might actually show. He had been actively soliciting sighting reports for his files since the Guthrie encounter in 1980. By the early 1990s, he boasted a file of over 200 eyewitness accounts, mostly from the middle part of the 1980s.[51] Burton lent the Enigma Project documents and information, to be sure, but he also lent the group his *reputation*. His sober and professional approach to the monster had laid the groundwork for the Enigma Project to study the Frew tape in a serious and scientific manner. Burton was by all accounts a self-effacing man, and his articles on Chessie invariably focused on everyone involved with the story except himself. Still, it must have been rewarding for him to see his fascinated skepticism pay off, at last, with a proper scientific investigation of this *thing* he had been following so closely for nearly two years.

By July 1982, the Frew film had caused a worldwide furor. From its humble beginnings in water ballet on the Potomac, Chessie had become a movie star, and the monster's greatest act was still to come. But its journey to stardom was indexed to some of the cultural rifts that existed in communities around the ring of the Chesapeake Bay, and also to questions about what Chesapeake identity was and who got to determine it. As the decade progressed, the answers to these questions would remain as indeterminate as the identity of the creature the Frews filmed. But the monster, unimpeded, would swim on.

Chapter 3

Dissecting the Frew Film

MICHAEL FRIZZELL AND ROBERT LAZZARA were very busy men in the summer of 1982. With a copy of the Frew tape now in hand, they made plans to contact scientists who could study it from both technical and zoological angles. The technical side of this process was easier to deal with. Although the pair were convinced from their meeting with the Frews and Rosiers that the film was not a fake, they still felt obligated to submit it for objective analysis. With their connections in the engineering community, it would be a cinch to find someone to do the job. The zoological dimension of things was a thornier prospect, although Frizzell and Lazzara probably did not realize just how difficult it would be. In order to identify what was in the Frew film, analysts had not only to untie the huge knot of assumptions that had grown up around the monster since 1978 but also to guard against imposing their own biases on the video while they were studying it. It was a lot to expect from four minutes of amateur footage, not to mention the people reviewing it.

As it is wont to do, the press was quick to make pronouncements about what the video contained. It seemed to take its lead from Karen Frew, who repeatedly insisted, as she did to the *Kent Island Bay Times*, that the creature was not "your run-of-the-mill water snake." But did that mean that she thought she had seen some kind of *mutant* snake? And if so, would such an animal constitute a "monster"? (An exasperated Frew also complained about being "quoted all over the country saying things I never said," so perhaps we will never know exactly what she thought she saw at the time.) The *Baltimore Sun* muddied the waters even further by asking, "Is the creature captured on the videotape Chessie, as Mrs. Frew believes? Or is it some species of snake that somehow found its way into the bay?" So was there a distinction between Chessie the monster and water snakes, regular-sized or otherwise? Even more confusing was the *Sun*'s assertion that the Frew creature was "amazingly similar" to what Trudy and Coleman Guthrie had seen in 1980.[1] Even a cursory reading of the Guthrie account pointed away from a serpentlike animal.

In order to get to the bottom of this mess, biologists would have to check their preconceptions at the door before evaluating the tape. In spite of the press's desire to oversimplify the matter, there were several interrelated determinations to be made before the question of serpents and monsters could even come up. To wit: (1) Did the tape actually show anything at all? (2) If so, was that object moving? (3) If so, was it moving under its own power or being pushed along by the current? (4) Mobile or not, was the object in the video an animal? If experts could decide even a *single* one of these questions, then *maybe* they could turn their attention to the identity of the thing in the water. Even then, the possibilities were almost endless, running the gamut from giant anacondas, to sea serpents (was there even a difference?), to some kind of outlandish cryptid like the plesiosaur—with all the commonplace possibilities like manatees, seals, and sturgeons to consider as well.

The permutations were so overwhelming, and the answers ultimately so open-ended, that it is no wonder everyone just went with the assumption that the video showed an exceptionally large snake of some kind, with the unspoken corollary that that is what Chessie was and had always been. Even Frizzell and Lazzara avoided tangling themselves up in the philosophical debate over both tape and monster. Frizzell adamantly told reporters that the creature in the video "definitely does not belong" in the bay, and 20 years later he steadfastly held that the video was the "hardest evidence" up to that point of a cryptid in the waterway.[2] This tendency—by everyone—to gloss over the conceptual nuances inherent in Chessie and the Frew tape is certainly understandable, but it did neither the monster nor the film any favors as they headed off to be subjected to close and potentially withering scientific scrutiny.

The Prescription for Monster Fever

Aided by the presumption of the Frews' innocence, the matter of evaluating their tape for a hoax was a speedy operation. Lazzara took the lead, bringing in television engineer Michael Rupert to examine the video on "special enhancement equipment." Rupert concluded that it was "virtually inconceivable that the tape is a fake." With the medium itself thus verified, Lazzara and Rupert turned their attention to its actual content. Was the object that the Frews filmed staged somehow, perhaps by being towed on a line? Was it even possible to know for sure? Here, some of the more existential problems of the Frew tape were coming to the fore. Nevertheless, after studying the video in extreme slow motion, Lazzara was convinced that the object was genuine. "There's not someone under the water pulling it along. There's not some line tied to it and someone pulling it," he told the *Baltimore Sun*. "It's alive."[3]

Now the Enigma Project turned to snake experts. The first port of call was Frank Groves, a herpetologist and then-curator of reptiles at the Baltimore Zoo. Groves was not convinced the video depicted a giant, tropical snake. "He shot that down real fast," Frizzell later recalled, "because of the Maryland climate, and the salinity of the water."[4] An anaconda would find the Chesapeake to be an inhospitable environment under even the best circumstances. With just one treatment, the Anaconda Syndrome was cured.

For a second opinion, Frizzell and Lazzara turned to George Zug, who was, at that time, chairman of the Vertebrate Zoology Department at the Smithsonian Institution's Museum of Natural History. Like Groves, Zug was an expert in herpetology, but he had also demonstrated an interest in, and sympathy for, cryptozoological matters by lending his expertise to investigations of the Loch Ness and Lake Champlain Monsters in the previous decade. Moreover, Zug was a founding member and director of the International Society of Cryptozoology (ISC), which formed in 1981 and held its initial meeting at the Smithsonian in January 1982. Zug described himself as a "skeptical believer" in the possibility of previously unknown, large lake animals. "Because of my willingness to examine photographs of unknown animals," he wryly wrote a class of seventh-graders in March 1981, "I have obtained the reputation as the Smithsonian's 'monster' expert."[5] Zug's cryptozoological reputation may have preceded him, but it was his scientific curiosity that caught the Enigma Project's attention—not to mention his geographic proximity to the Chessie phenomenon. Frizzell figured that "if Zug was interested in Loch Ness, then he may be interested in something that's a lot closer to home."[6] Frizzell approached Zug about hosting a small conference so that he and other experts could analyze the tape for themselves. Zug was open to the proposal, so a panel was scheduled for mid-August.

The cryptozoological community greeted news of the conference with exultation. Chessie, along with Champ, the Lake Champlain Monster, brought new legitimacy to the emerging science of "hidden animals" and especially highlighted the need for closer investigation of the many American and international waterways supposedly harboring unknown aquatic creatures. Chessie and the ISC were products of the same 1970s zeitgeist, so it should come as little surprise that they emerged around the same time, and that their stories became intertwined. What *is* a little surprising, perhaps, is the speed at which Chessie became a major focus for cryptozoologists. Chessie quickly outswam its fellow sea serpents for two main reasons. First, by 1982, multiple expeditions to Loch Ness had turned up few pictures, even less hard evidence, and certainly no Nessie carcasses to study. Much the same was true for Nessie's well-documented American sister, Champ. Second, and by contrast, evidence for Chessie seemed more plentiful and of a higher quality, especially given the short time since the monster had started making appearances. Even Chessie's sighters seemed to be more expert than the stereotypical eyewitness. Chessie became the darling of cryptozoologists because, for a brief few years, the monster seemed like the best hope for proving the legitimacy of cryptids—and also proving the legitimacy of cryptozoologists.

In only its second issue, the *ISC Newsletter* featured a story on Chessie and the Frew video that connected the breaking news from the Chesapeake with recent "worldwide interest" in the Champ phenomenon. "Chessie is not new to cryptozoologists, nor to the folklore of the Chesapeake Bay/Potomac River region," the article asserted. "Sighting reports go back decades, and are similar to reports of animals in several other marine regions of the world." It was a *very* liberal interpretation of scant evidence, but the hint of a deeper history for Chessie legitimized the monster and opened the door for stars in the cryptozoological firmament to weigh in on the creature's identity. ISC vice president Roy Mackal, a University of Chicago biologist renowned for his expeditions in search of the Loch Ness Monster and the putative Congolese sauropod Mokele-Mbembe, suggested that Chessie was a zeuglodon—an ancient ancestor of modern whales resembling a serpent. Forrest G. Wood, a marine biologist and ISC board member, argued just the opposite after seeing the Frew tape (presumably on the news). "I'll say with considerable certainty that it wasn't a cetacean," he wrote newsletter editor J. Richard Greenwell. Greenwell floated other possibilities, based on the work of Bernard Heuvelmans, the father of cryptozoology and ISC president, whose *Le grand serpent-de-mer* (1965; first published in English as *In the Wake of the Sea Serpents* in 1968) established a taxonomy for unknown sea creatures.

Heuvelmans's system of categorization grouped together sighting reports according to physical or behavioral similarities noted in the creatures by eyewitnesses. In turn, those similarities were categorized as "types" by which creatures in future encounters might be identified. On the basis of witness descriptions, Greenwell suggested three categories to which Chessie potentially belonged: "long-necked" (that is, plesiosaur-like), "super-otter" (unusually large marine mammal), or "many-humped" (the classic sea serpent). What Chessie could not be was a "super-eel," Greenwell argued, because eels only reach about four feet in length in the Chesapeake. Greenwell also offered the anaconda theory, although he confessed that it was uncertain how anacondas could survive outside the tropics. In the end, he philosophically concluded, "all the Smithsonian can do is try" to identify Chessie. By the 1980s, *In the Wake of the Sea Serpents* had become one of the holy texts of cryptozoology, so it is no surprise that Greenwell referred to it to put Chessie in context.

But there was a certain amount of futility in trying to cram the monster into the Heuvelmans taxonomy—itself an ambiguous enterprise—when no two Chessie sightings were exactly alike in the first place.[7]

The five weeks that passed between July 11, when the *Baltimore Sun* broke the news of the Frew tape, and the Smithsonian meeting, on August 20, proved to be a trying time for all involved in the Chessie phenomenon. The main source of everyone's misery was the media. Although the Eastern Shore press had largely taken the tape in stride, the *Sun* article sent the story into the stratosphere, and national and international outlets quickly descended on the Frews, Zug, and anyone else associated with the developing story of the film. Stirred up by the *Baltimore Sun*'s story about the tape on July 11, the media became so intrusive that Zug withdrew completely from the conference planned for August. As the *Richmond Times-Dispatch* put it, he had "received so many phone calls about Chessie that he has decided to forget the study and return to the silence and solitude of his laboratory." Zug's own take on the decision was characteristically direct. "It's been a media circus, and we're about 100 percent sure we're going to disassociate ourselves from it," he told the press flatly.[8]

The Frews were equally irritated by the media's antics. "Every television station and newspaper imaginable has called," Robert Frew complained. At the same time, however, Zug's abrupt resignation from the planned panel also caused consternation. The couple hoped that after the initial furor died down, Zug would still go forward with the conference. "The fact still remains," Karen Frew remarked to the *Star-Democrat*, "that it is his field and if he's half as interested in this as we are, I feel confident he'll do it."[9] The Frews' disappointment with the scientific community was compounded when the National Geographic Society made it known, at about the same time Zug withdrew, that it had no interest in the video. Robert Frew had approached the society personally, hoping to sell the video, or at least to seek the society's scientific opinion. Instead, Emory Kristoff, an underwater photographer for the society, offered withering criticism: "It looks like four kids swimming inside a plastic bag," he told an interviewer. "It's so jerky and amateurish . . . you can't tell very much from it."[10] Frew *did* catch the attention of the Virginia Institute of Marine Science, whose fisheries director told the *Times-Dispatch* it would be glad to examine the tape. But the institute was a little too lacking in prestige now that Chessie was a proper film star, and Chessie's Potomac prehistory was virtually forgotten at this point anyway. If the offer ever reached the Frews, it was never seriously considered.[11]

The explosion of public interest in Chessie went beyond newspaper sensationalism, especially once Zug was connected with a potential investigation. The zoologist apparently received a mountain of letters from private citizens concerning the monster, most of which seem to have been forwarded to Bill Burton for his files. A very small selection of these letters survives in the Smithsonian archives. By far the most intriguing came from Joe Perdue, a newspaperman from Nellysford, Virginia, who wrote on July 12, 1982, to say that he believed he had captured photographs of something similar to the Frew creature near the mouth of the York River (about 60 miles south of the mouth of the Potomac) in summer 1980. Perdue sent three color negatives to Zug for study, but the latter was unable to determine what animal they showed, if any at all.[12]

The Chessie Advantage

Although the *Times-Dispatch* asserted that the Frews had never sought the publicity that Chessie brought them, the claim was a little disingenuous. After all, Frew himself had made the first crucial contacts with the media that had led to the explosion of publicity around his tape, and he had also made overtures to the

National Geographic Society. Frew also candidly admitted that he hoped to cash in on Chessie. "You can't tell me the person who took pictures of Big Foot didn't make more than 50 cents from it," he told the *Star-Democrat*, which reported that Frew was then in the process of securing the copyright to the video before it became widely distributed. According to Frizzell, who had advised Frew to ensure that he retained ownership of the video and its copyright, that process would prove even more frustrating than instant fame, and was ultimately unsuccessful. News stations apparently took advantage of the Frews, making copies of the tape and refusing to compensate the couple.[13]

It was not just Robert Frew who saw dollar signs because of publicity generated by the tape. Chessie had always flirted with the tourism industry and had even, because of its summer appearances, been likened to a tourist. But it was not really until 1982 that Chessie turned the corner from a local oddity to a potential tourism *asset*. The *Star-Democrat* broached the idea in a semicomical editorial on July 1. Boasting of the "Chessie advantage," the paper concluded that the film had "two immediate implications for the Eastern Shore, one of them bad, the other good." The "bad" implication was that Chessie "poses a definite threat to swimmers, watermen and recreational boaters who during the warm weather thrive on the Bay and its resources." Fear of the monster might, therefore, cause the region to "lose the benefits of tourists who come from elsewhere to take advantage of the Bay and who would likely take their families and boats elsewhere for the weekend or the vacation." To prevent this "disaster," the paper called for the employment of "every available surveillance technique," including analysis of the Frew film, to identify the creature and determine whether it was dangerous. The editorial suggested the monster should be "preserved for the sake of natural history" if possible, or—if carnivorous—a reward should be posted for its capture and extermination. Breaking with the traditional depiction of Chessie as a "gentle" monster, the column lapsed into a joke that resurrected some old themes: the whole cast of *Jaws* (1975) should be reunited for such an expedition, the paper claimed. "That brings us to the favorable implications of this Chessie chapter," the piece concluded: "the national exposure the Eastern Shore and Chesapeake Bay would receive" when *Jaws*-knockoff films starring Chessie came to theaters. The Shore's tourism promoters would have their work cut out for them, but it would be worth it. "If the Shore really wants to put itself on the map, it should milk this Chessie thing for all it's worth.[14]

Despite its light tone, the editorial was as much a snapshot of the phenomenon surrounding Chessie as the Frew video purported to be of the animal itself. It not only wove together themes that had underpinned sightings since 1978 but also hinted at the shape Chessie was beginning to take in 1980s Maryland. This new, emerging Chessie represented the triumph of suburbanization and tourism, as well as the opening of the Eastern Shore of Maryland—a region not unlike Virginia's Northern Neck in its culture and economy—to the outside world. Chessie was becoming a symbol: an ambivalent one for now, to be sure, but one full of potential, swimming in increasingly fecund waters. "Psst! Did you know this about the Shore?" winked a "guidebook" supplement the *Star-Democrat* published the same day as its editorial. "Chessie is not the nickname for a well-know [*sic*] railroad. It's the Chesapeake Bay's 'Loch Ness' monster, seen by a handful of sober, and otherwise well-respected, watermen and pleasure boaters as it cruises placid waters."[15]

Placid waters they might have been, but the media remained determined to make waves. As the summer of 1982 wore on, the Frews fielded hundreds of phone calls from news agencies around the world, at all hours of the day and night. *CBS Evening News* anchorman Dan Rather even called. Television vans and cars lined the narrow road to their home, blocking traffic and making travel difficult for the Frews and their neighbors. The barrage quickly began to take its toll on the family. "I'm pretty exhausted," Karen Frew confessed to the *Star-Democrat*.[16]

The Frews also faced increasing pushback from naysayers. One man, John Masone, came forward in mid-July to assert that the video showed a family of otters swimming in a line. Masone claimed he had encountered the same phenomenon himself several years before. "What I saw was very long and it looked like a big snake," he told the *Star-Democrat*. "It scared the pants off me, but I cornered it and it went into the rocks. Just then a mother otter jumped on the rocks and started hissing at me." The Frews took umbrage at Masone's claim. According to the newspaper, they were familiar with what otters swimming in a line looked like "and would know the difference between that sight and what they claim to have seen in May." "It's definitely not the same thing," Karen Frew snapped. "He wasn't here and it has been determined to be only one thing out here."[17]

Perhaps the worst lashing the Frews received came from *Baltimore Sun* columnist James Holechek, who wrote bluntly that "these so-called phenomena are either misinterpretations by the observers, sensationalistic reporting or fraudulent acts." Although he conceded that sighters could be "honest and well intentioned," Holechek nonetheless concluded that there were many natural, plausible explanations for Chessie that need not involve sea serpents. For instance, he asserted that the Guthrie sighting in 1980 was probably a manatee. His observations about the Frew film are worth quoting in full:

> Like the Frews, I live on the bay and look out over its surface in all kinds of weather. Just when you think you've seen all of its characteristics, something new occurs. I've seen one wave rise mysteriously 2 feet over a flat calm surface, then speed toward the shore. A ship could cause such a wave, but in several cases, none was

> in sight. Old-timers acknowledge the strange waves, but none can explain them. Ship and boat wakes do look like the backbone of a spiny snake, especially at certain times of day. The waves move in one direction, sometimes disappearing then reappearing, sometimes shortening or lengthening.

Holechek went on to propose that the Frews could also have misidentified tubing used to contain oil spills; he had encountered some of this himself on a recent boating excursion, directly off Love Point. Whatever Chessie was, however, Holechek was prepared to argue that it was not a sea monster.[18]

Although it had been implicit in the sightings since 1978, the videotape brought to the surface, as it were, how much Americans depend on visuality, almost exclusively, as a basis for gathering and evaluating knowledge and creating culture. As Chessie amply demonstrates, far from revealing objective "truth," sight is subjective and unreliable at the best of times, and things that are seen are shaped by the personal prejudices of their observers.[19] As cryptozoologists Loren Coleman and Patrick Huyghe have noted, "From a distance, just about anything might look like a Lake Monster or Sea Serpent. And then there is psychological contagion—if people are told to look for monsters, they will see them."[20] In Chessie's story, not only were people encouraged to see sea monsters, they were also encouraged to believe that merely looking at the water constituted some kind of knowledge of it. When people who *really did* know the water—fisheries officials or watermen—denied the existence of the monster, their opinions were regarded as suspect somehow or just ignored completely.

The video especially enabled Chessie to bypass those troublesome watermen and their annoying skepticism. Seeing *was* believing: there was no need to consult watermen for their opinion on Chessie when the evidence was plainly visible in living color on a television screen. Watermen still mattered, though: *only* they could provide the local wisdom and salty stories that rooted the monster in bay lore. Accordingly, the media changed strategy, co-opting watermen by implying that they welcomed the video as much as the general public. The *Bay Times* eagerly reported Robert Frew's tale that "older watermen in the area that have heard about the sighting have very little reaction." He claimed that they responded, matter-of-factly, "Oh yeah, we know something's out there."[21] Another report suggested that watermen found the film a relief: now they could admit what they had always been seeing but had to keep to themselves out of fear for their reputations. Karen Frew claimed that "Eastern Shore watermen came up to Robert Frew to shake his hand. 'Over at Tilghman Island,' she said, 'the captains were coming up to him and saying they were glad somebody finally got some proof. . . . They had Bob sign napkins saying that they could have seen Chessie because Bob took pictures of it.'" The *Baltimore Evening Sun* even went so far as to call Robert Frew "a celebrity among Eastern Shore watermen."[22] In a time and place where watermen were increasingly romanticized, even their skepticism was a commodity to be packaged and sold.

Creature Feature

Despite George Zug's reluctance to tilt with the media, he eventually relented and agreed to host the Chessie panel in August after all. Breaking the news on July 15, the papers could barely contain their excitement. Smithsonian spokesman Tom Harney told the Associated Press that Zug would be examining the tape to see if Chessie's "behavior or posture resembles any known animal." The results of the panel would be issued in the form of a press release. There would be no press conference afterward because Zug was gun-shy of reporters after having been hounded by them at a recent conference devoted to the Lake Champlain Monster.[23]

In the month preceding the conference, the Enigma Project received several Chessie sighting accounts, some of which came from what Robert Lazzara called "closet witnesses," people who, like the putative watermen, had seen the monster but had not come forward out of fear for their reputations.[24] Among these witnesses were Carol Taylor and her father, Clyde, who claimed to have sighted the monster from Cloverfield Community Beach, not far from Love Point, on July 16, 1982. A commercial artist, Clyde Taylor subsequently made a set of detailed drawings of the monster for the Enigma Project. The Taylors' encounter was a little different from the norm, in that they allegedly came upon the sea serpent while it was basking in the shallows and got quite close to it. Their description was unusually meticulous and included even the color of its scales.[25] Although atypical, the Taylor sighting gained a small measure of fame in its own right when it was featured on a 1986 BBC Radio program on sea monsters, and again in a 1994 installment of the *In Search Of . . .* knockoff series *Mysteries, Magic, and Miracles*, hosted by former *Avengers* star Patrick Macnee.

Another encounter that went public at this time came from Kathryn Pennington, of Ellicott City, Maryland. The *Baltimore Sun* reported on July 18 that Pennington had not just seen but photographed the creature while visiting Goose Point, on the Choptank River, on May 24, 1981—a year before the Frew video was taken. She had not come forward at the time out of fear of ridicule, but the press surrounding the video had convinced her to do so now. Pennington's description of the creature she saw matched that of the typical Chessie sighting. Her blurry snapshots, on the other hand, showed a "squiggly figure in water with a shoreline in the background," as the *Sun* dryly put it, as well as what appeared to be the head of the creature rising up out of some grass on the beach. The paper's disappointment was almost palpable when it told readers the images were of insufficient quality to be printed in the paper. As they were the only other extant visual evidence of Chessie besides the Frew tape, however, the Enigma Project asked Pennington to submit them for the forthcoming Smithsonian panel.[26]

The Smithsonian meeting was set for 11:00 a.m. on Friday, August 20, at the Natural History Museum's Naturalist Center. Zug and the Enigma Project composed separate guest lists, with the former enlisting half a dozen Smithsonian colleagues, as follows: Frank Ferrari and Leslie Knapp, of the Smithsonian Oceanographic Sorting Center; Nicholas Hotton III, from the Department of Paleobiology; James Mead and Charles Potter, of the Division of Mammals, Department of Vertebrate Biology; and Clyde Roper, chairman of the Department of Invertebrate Zoology.[27] For its part, the Enigma Project invited several regional fisheries authorities, scientists from the National Aquarium in Baltimore, the Frews and Rosiers, Pennington, and a few other interested parties. In all, the guest list totaled about 30 people. True to Zug's instructions, no one from the media received an invitation, with the exception of Bill Burton, whose presence Zug permitted as a favor to the Enigma Project because of Burton's reputation and "extreme vested interest" in Chessie.[28] Frizzell and Lazzara were thrilled that so many specialists would be bringing their expertise to bear on the Chessie phenomenon. Writing to Zug to submit their list and thank the Smithsonian for hosting the meeting, Frizzell affirmed the Enigma Project's view "that the participation of you and your colleagues, as well as other invited professionals, will provide an excellent cross-section of scientific and authoritative disciplines by which the Frew video tape and the Pennington photographs can be scientifically and objectively examined."[29]

The day of the conference, participants were shown the Frew video and three enlarged prints of the Pennington photographs, and "a very open and frank discussion," as Zug later put it in a memorandum of the event, ensued. Because it was moving footage, the Frew tape was the focal point of the deliberations, with the experts closely examining every detail of the object's shape and movement, and the conditions in which it had been filmed. According to Zug, little of the object's physical makeup could be discerned based on the video. Its "head"—if that is what it was—had no apparent facial features, like eyes or a mouth, and the shadow behind the "head," which had been taken to be the extended tail of a serpentlike creature, might just have been a wake. Zug did concede a slight similarity of shape between the "heads" of the Frew and Pennington objects, but otherwise no conclusions could be drawn about either of their identities. One conclusion the conference *did* reach was that the object in the Frew film was not undulating vertically, as the Frews and many eyewitnesses before them had reported of Chessie, and may not even have been doing so horizontally. It *was* moving, Zug wrote, "but the manner of propulsion could not be determined." At the end of the day, consensus had been reached on one point alone: "All the viewers of the tape came away with a strong impression of an animate object."[30]

A carefully worded Smithsonian press release delivered the meeting's verdict with clinical detachment:

> In spite of the eyewitness description and the visual evidence, no identification, not even a tentative one, was offered for the strange sightings. There was not enough visible evidence on the tape for a positive identification. The usual explanations of a partially submerged log, a string of birds or marine mammals, optical illusions, etc. seem inappropriate for the dark, elongated, animate object. The Chessie phenomenon differs in many ways from the typical sightings of marine animals. The front end is strongly angular, appearing and disappearing vertically instead of rolling above and below the surface as

> in most swimming animals. Similarly, the object undulates in a vertical plane, according to eyewitnesses, rather than a horizontal plane as is the normal swimming movement for elongated animals. The photographs and videotape are intriguing and identification of the object or objects in these films might be aided by some form of image enhancement.[31]

The Enigma Project hailed the announcement as a triumph for cryptozoology. "The nature of that press release is terribly encouraging as to the continued study in looking for an unexplained animal in Chesapeake Bay," Frizzell told one newspaper.[32] Thanking Zug for his help with the meeting in a letter on August 27, Frizzell applauded the Smithsonian for giving the video and photo evidence "serious scientific consideration in view of its highly controversial nature." For Frizzell, the conference was a watershed moment for the legitimate scientific study of the paranormal. "I believe we are entering an age of incipient awareness in that reports of such natural phenomena, as Chessie, are no longer 'swept under the rug' simply because they challenge or deviate from established scientific doctrine," he enthused.[33]

Other attendees were less exuberant. In a lengthy letter to the *ISC Newsletter* in late 1982, Craig Phillips, then director of the National Aquarium in Baltimore, stated his view that the object in the film was "some kind of living creature," but probably not an animal unknown to science, and possibly not even one with the serpentine shape that it appeared to have. "Seen at a low angle and at a distance, almost any submerged object may appear as an elongated shadow at first," he wrote. One candidate for the object's identity that Phillips roundly rejected was an anaconda: such snakes do not deviate from shallow water and, in any event, would either starve or freeze to death in the Chesapeake's mid-Atlantic climate.[34] A blunter assessment comes from Nick Carter, a Maryland fisheries scientist whom the Enigma Project invited to the conference. Writing in 2018, Carter recalled that conferees were shown "a pretty blurry film" and that he "was not impressed or excited by the possibility that it was either something alive, or that the film was something that would help clarify anything about 'Chessie.'"[35] A ringing endorsement of paranormal research the conference was not.

Even Frizzell was more underwhelmed by the panel's conclusions than he let on. "I think deep down inside, I knew that it wasn't really going to go anywhere," he remembers. "But . . . it was a foot in the door—that's how I looked at it: 'OK, it probably won't go anywhere but at least they're sitting still long enough to look at it,'" he reasoned. "And to me that was a big achievement for the Enigma Project, and for cryptozoology, and I wish that it had gone further."[36] Truthfully, Zug himself was probably as disappointed as everyone else by the conference's outcome. "It was a fascinating, but frustrating, experience," he wrote in a form letter reply sent out in response to public inquiries about Chessie.[37] According to Frizzell, Zug told him privately that the Frew tape was "something very interesting."[38] Perhaps he wished he could have been less bound by his scientific principles and methodology, and more definitive about what the film and photos had shown.

For its part, the media—already superheated into a frenzy by the possibility that there might be actual proof of a sea monster in the bay—was left totally bemused by the Smithsonian press release. Had the conference proved or disproved Chessie's existence? No one could tell for sure, and that kind of ambiguity was bad for the news business. For the cryptozoological community, however, the press release was a revelation and a gift. It "sounds like it plays down the significance of the film," Lazzara told the *Sun*, but by *not* pronouncing on the creature's existence, it left the door open for further study and discussion.[39] Interviewed in October 1982, Lazzara reiterated that the Smithsonian had validated the existence of "a genuine, unknown phenomenon swimming the Chesapeake Bay's waters." "This is important because up to the Frew videotape, Chessie was nothing more than local legend," Lazzara said.[40]

Times-Dispatch wag Charles McDowell—who had always treated Chessie as a comic figure and figment of the newspaper silly season—echoed these thoughts, if ironically, in a column in late August. In a fictional conversation with his aunt Gertrude from rural Virginia, McDowell discussed the "real news," the "biggest thing that's happened up there [in Washington, DC] this summer." Gertrude chastised McDowell for failing to report on the Smithsonian press release. "What we have here," "she" said, "is a serious scientific judgment that the thing cannot be dismissed as an illusion or a hoax. That's news. . . . Get that, coming from the Smithsonian's Museum of Natural History, with all the credible overtones thereof."[41]

Aunt Gertrude must have belonged to the ISC, because that group's leadership was also clamoring for more news about the conference. In early September, J. Richard Greenwell wrote Zug asking for a report. "What's happening with Chessie? People are asking me about the analysis, but there is nothing I can tell them. . . . What were your conclusions, if any?" An apologetic Zug promised Greenwell a reply by the end of the month, but it was not until spring 1983 that an article appeared in the *ISC Newsletter*. Largely a reiteration of the press release and his own memoir of the event, Zug's article stuck to the facts and avoided undue speculation. A final statement, likely inserted by Greenwell, nevertheless almost perversely struggled to keep the cryptozoological faith alive: "Like its cousins in other parts of the world, Chessie remains both an 'unexpected' and a 'cryptic' animal, despite the opinion of some of the Smithsonian's leading biological scientists that the videotape, at least, probably depicts an animate object."[42] Considering that the tape had once been regarded as the most convincing evidence yet for the existence of Chessie—giving hope to monster hunters everywhere—Zug's article brought the Smithsonian episode unceremoniously back down to earth.

There was worse to come for Chessie, though. Writing in late 1983 to a different cryptozoological newsletter, Frizzell declared that attendees at the Smithsonian conference had "unanimously concluded that the tape depicted a long 'definitely continuous' animate object." A subscriber to that newsletter also, Zug mercilessly dismantled Frizzell's assertion. "I have yet to discuss a topic with a group of my colleagues and obtain unanimity on the interpretation of any observation," Zug retorted. "Cryptozoological observations certainly do not attract agreement among observers let alone those trained to be skeptics." The only consensus among conferees, Zug added, was that the video *probably* showed an animate object.[43] Any other conclusions were a matter of conjecture.

The Man with the Midas (Re)touch

If there was one thing all parties could agree on in the wake of the Smithsonian conference, it was the proposition that image enhancement might reveal something about Chessie's nature that the experts had missed with the naked eye. It was probably an overly optimistic hope that was rooted in the 1980s fascination with emerging computer technology, and it also extended the theme of hypervisuality that had helped give rise to the Chessie phenomenon in the first place. But the possibility of image enhancement gave the press something monster related to hang on to after the anticlimax of the Smithsonian conference, and the prospect quickly became the focus of public interest in the monster.

The notion that the Frew video would yield more secrets if subjected to computer analysis was suggested by Zug at the time of the conference. Zug had contacts at the Jet Propulsion Laboratory (JPL) in Pasadena, California, and offered to connect the Enigma Project with them. Around the same time, Andrew Goldfinger, a staff physicist at the Applied Physics Laboratory (APL) at Johns Hopkins University, approached Frizzell about doing the enhancement pro bono. "Although I fully intended to pursue Zug's suggestion," Frizzell later wrote, "I soon found that it was apparently pursuing me."[44] APL and JPL were competitors in the emerging field of computer image enhancement, so when news broke that the Frew tape might be sent to Pasadena, APL offered its services "in an attempt," Frizzell observed, "to improve its public image and to avoid being outdone." Interest from *two* such prominent institutions was a real embarrassment of riches, and Frizzell knew it. "Between JPL and APL the Enigma Project will surely have the videotape enhanced," he told Zug a week after the conference.[45]

In the fall of 1982, optimism ran high that APL would quickly unravel Chessie's secrets. "The computer image enhancement could do one of two things," Frizzell giddily told the *Baltimore Evening Sun* in late August. "It could give us evidence to go on with our investigation, or it could tell what the Frews spotted in the bay." By mid-September, Frizzell had evidently accepted Goldfinger's offer, and a copy of the Frew tape was handed over to APL. Frizzell hoped the work would be completed by the end of the month.[46] As it turned out, the process took longer—a good deal longer.

It took *two years* for news of APL's progress to become public, and the results, when they finally came, were unfortunately underwhelming. The enhancement process, Russ Robinson wrote for the *Baltimore Sun* in August 1984, "has convinced researchers that there is a live, unidentified creature prowling the waters of the Chesapeake Bay. But scientists say that even with computer-enhanced images they can't identify the creature from the videotape."[47] If this assessment seems familiar, it is because it strongly echoed the conclusions reached by the Smithsonian conference two years earlier. If the object of the exercise was to determine what Chessie actually was, it fell far short of expectations.

Maybe expectations were the problem. Computers so captured the public imagination in the early 1980s that they were seen as devices capable of working miracles. No doubt their introduction to the ongoing Chessie debate raised hopes about the possibilities of image enhancement. The term itself—"image enhancement"—conjured up a process more glamorous, and a product more definitive, than APL or anyone was capable of undertaking. What Goldfinger and his colleagues *actually* did, as opposed to what the public envisioned, was use computers to colorize and retouch frames of the Frew film in order to throw the object in them into sharper relief against its background. It was a painstaking process. As Goldfinger explained, "We have to be careful that we don't manipulate the computer to get the picture we want instead of the picture that is there." By toggling between enhanced frames in a sequence, Goldfinger could demonstrate that the object moved—and possibly suggest by what method—but that was about *all* he could do. Although high-tech for the time, it was a far cry from the science fiction wonders that computers otherwise seemed to promise in the era, and it failed to produce images that fired the public imagination. It is no wonder the news left everyone a little deflated. The dismay was palpable in the *Sun*'s headline: "Elusive 'Chessie' Stumps Computer."[48]

Goldfinger felt confident that further enhancement of the film could reveal more information about the creature's size, shape, and speed. Unfortunately, funding and time ran out before further work could be completed. Goldfinger estimated that it would cost upwards of $40,000 (in 1980s dollars) to complete the project—a substantial outlay for a niche project that he described to an interviewer in 1993 as "a rather low priority" taken on as "just a lark."[49] Ultimately, after two years of anticipation, the only tangible result of the APL enhancement was the introduction of a new moniker for Chessie: the "Chesapeake Bay Phenomenon." Goldfinger and his colleagues adopted the more scientific term, according to Frizzell, because they "felt [it] was more befitting the alleged beast."[50] Ironically, in contrast to the enhanced images the APL team produced, Chessie's new name put the focus on the bay rather than the creature. The background had become the foreground.

The APL story sank virtually without a trace shortly after it broke, and although the enhancement debacle was mostly benign, in retrospect it was probably the first in a long line of dents in the monster's reputation that led to its eventual demise. For a brief moment, Chessie had seemed to be a rare example of an explainable mystery that *gained* rather than lost mystique by its solution. Identifying Chessie would have only added to the allure of the Chesapeake, even if the monster turned out to be something totally prosaic. Instead, the ambiguity and circularity of Goldfinger's findings, coupled with the inconclusive Smithsonian conference, seemed to disappoint the public and undermine the goodwill the monster had enjoyed since 1980.

Once hailed as hard evidence for an unknown creature in the bay and a moment of increased legitimacy for cryptozoology, the Frew tape was now somehow diminished. "Scientists studying the tape at the John [*sic*] Hopkins Applied Physics Laboratory have never been able to decide what it [Chessie] looks like," scoffed William Rodgers of Centreville, Maryland, in an acerbic letter to the Easton *Star-Democrat* on August 17, 1984. With its poor sketches and inconsistent eyewitness descriptions, Rodgers was forced to lump Chessie in with other tabloid topics like UFOs and alien abductions. "Perhaps the *Star-Democrat* will continue to enlighten us about the Bermuda Triangle, the Amityville Horror, sightings of Big Foot, those visits from outer space, and that meeting Oral Roberts had with Jesus who stood seven stories tall while advising the faith healer to solicit funds for a vast construction project. This is great dog days material."[51] After the credibility Chessie had gained since 1982, to be relegated again to the silly season was a significant step backward from which the monster would never quite recover.

Chapter 4

A Serpent in Eden

WITHOUT INSTANT RESULTS from the video analysis to follow up on the Smithsonian conference, the Chessie story hit a lull. By the end of 1982, Chessie had strayed off the front pages and seemed to have submerged for the season, if not for good. The next year then proved poor for further manifestations. A charter boat captain insisted he saw the monster near the Woodrow Wilson Memorial Bridge in the Potomac in April (experts countered that it was debris from a recent flood), and a whale apparently swam into waters adjacent to Kent Island in June, but otherwise reports of the monster were few and far between in 1983.

Still, there was an almost unconscious demand for Chessie. In their style and language, the whale-sighting reports consciously mimicked those of monster encounters: the sighters, a family of three women—two local, one a visitor—were sitting on a community beach; water-skiers seemed to spot the animal and exited the water shortly afterward; scientific experts pronounced thirdhand on the credibility of the creature's presence in the bay. All the hallmarks were there for a classic sighting report. What prevented the whale's classification as a "monster" was the fact that it had breached as it swam by its observers, seemingly a clear sign that the animal was a cetacean, or at least a known animal of some sort. Yet the encounter raises the question, If the whale had *not* breached in front of them, would eyewitnesses have reported seeing Chessie instead?[1]

Near misses like the one in the Potomac, or even the whale sighting, peeved the press, which had helped make Chessie a ubiquitous presence and had a small but vested interest in keeping the creature around. "The day is past when any sighting could be treated like almost any other sighting," *Richmond Times-Dispatch* wit Charles McDowell wryly observed in July 1983. "Now Chessie spotters have a new standing in the world, a new dignity in the eyes of science and must strive always to be worthy of it."[2] McDowell was joking—mostly—but he was also voicing an important truth about Chessie's status after the Frew film. If the monster were going to compete with other big-league cryptids, its eyewitness accounts had to be of pristine quality, and plentiful. Whatever the quality of Chessie reports—and they were, as we have seen, pretty variable—one thing they were not, as a general rule, was abundant. Its patchy sighting record is probably one of the reasons Daniel Cohen, author of *The Encyclopedia of Monsters*, disdainfully told the *New York Times* in 1984, "In the world of monsters, Chessie is not an important one. It doesn't rank very high."[3]

Regardless of how it compared to its peers in other regions, Chessie was important in the Chesapeake, especially on the Maryland side, and growing more prominent with every new sighting, even if those were still intermittent. Why were people so determined to believe the monster existed, even when the proof for it—the Frew film included—was nebulous at best?

Here we return to those larger questions of time and place that we have been deferring. Why did Chessie appear when and where it did? And why was Kent Island, a place filled to the brim with skeptical watermen, seemingly the epicenter of so much Chessie activity thus far—and much more to come?

To answer these questions, it is necessary to understand what was happening on, and *to*, Kent Island in the early 1980s. Like the Northern Neck of Virginia and many other communities in the ring around the bay, the island was a battleground for two competing visions of water, as suburbanization transformed the waterfront and altered both the cultural and physical fabric of the region. *Unlike* the Northern Neck, where the onslaught of development was blunted by its relative seclusion, Kent Island was easily accessible to outsiders from urban Maryland and the nation's capital. By the time tourists and retirees along the Potomac were just *beginning* to fret about a monster being in the water there, something much more threatening was already in the process of swallowing up Kent Island, hook, line, and sinker. In Virginia, Chessie was a harbinger of things to come. In Maryland, it was a symptom of how far gone things really were.

One Fish, Two Fish, Red Fish . . .

Although 1983 had been a disappointing year for sightings, conventional wisdom held that Chessie only appeared in even years anyway—so if a manifestation were going to occur, 1984 seemed like the right time for it. The notion that Chessie avoided appearing in odd years was a convenient bit of lore invented in 1981 in order to "explain" why the monster had only been seen in 1978 and 1980. *Times-Dispatch* writer Albert Oetgen expressed the idea as if it were a commonly held view in a December 1981 article, so presumably jilted Chessie hunters had already adopted this rationale by then.[4]

Over in Maryland, another theory emerged. Bill Burton's years spent compiling Chessie sightings led him to believe that the creature was following bluefish runs. A migratory species, bluefish swim northward in the summer; on the Atlantic seaboard, their route takes them into the Chesapeake Bay and its tributaries. Burton proposed that in years of lean bluefish runs in the Chesapeake, Chessie's visits to the estuary had also been reduced or nonexistent. The theory also seemed to account for the creature's appearances in vastly different areas of the bay across a given year.[5]

Not surprisingly, Burton's hypothesis was vague as to *why* Chessie would follow the bluefish. To eat them? Because it *was* a bluefish? Because it just liked company on its travels? Its haziness aside, the bluefish theory caught on quickly. By the fall of 1982, it had become accepted as part of the lore that, "according to the legend, Chessie usually shows up on their tails."[6] What it lacked in provability, it made up for with a strong dose of "old salt" water knowledge, the kind of insight that only a waterman, or a proxy like Burton, would have. The theory was also just a little creepy. Consider Burton's account of a friend's fishing trip, which resulted in a Chessie sighting: "Just before it appeared, they had all kinds of bluefish activity. Bluefish break on the surface often. Before they even saw the creature all of the fish stopped breaking and it was very eerie, they said."[7] The language harked back to the "uncanny nature" theme that had been prevalent in the 1970s monster reports, with a subtle difference. Although Burton probably believed the bluefish theory made Chessie more explainable and scientific, and perhaps a more legitimate object of his and the public's interest, it had the effect of making the monster a little aloof and mystical.

For the Chessie faithful, however, Burton's proposition checked all the right boxes, and by the summer of 1984 the "bluefish theory" had replaced the "even-year theory" as the explanation for Chessie's intermittent manifestations. It was just as well, since by the end of July, it was looking as though an *even* year was going to be a bust for the first time since 1978. Only one report of the monster had come in by that time, down in the southern bay, and Maryland-based researchers seemed unenthusiastic about it. Interviewed in August, Michael Frizzell invoked the bluefish theory to explain the lack of sightings: "In 1982, there was a good run in the bay and numerous reported sightings. This year, bluefish are scarce and so are sightings."[8]

As it turned out, however, 1984 was *not* a lean a year for Chessie. On August 7, the day after Frizzell's observations appeared in the *Star-Democrat*, the newspaper announced a new sighting. Harry R. Lohman, a "semi-retired Kent Island builder," and his wife were coming home from a pleasure jaunt in the Wye River in the late afternoon when Mrs. Lohman saw what she thought was a log floating in the water near the Miles River. "But it was moving," she told the *Bay Times*. The Lohmans changed heading in order to get closer to the object. What they saw were three "bluish-black" humps projecting out of the water, "perfectly aligned and . . . moving in a pattern," speeding through the water faster than the Lohmans were. "And the part that was out of the water was longer than our 24-foot boat!" Mrs. Lohman exclaimed. When the Lohmans got to within about 30 feet of the creature, it dove. "Then we heard a loud thump," Mrs. Lohman told the *Bay Times*. "We think it may have hit the boat as it dove under, something hit us." The Lohmans immediately contacted the Maryland Natural Resources Police about their sighting, but, perhaps predictably, the agency had little to say on the matter. Although she confirmed it was tracking sightings, spokesperson Lenett Davis added, unhelpfully, "I can't say whether the Department of Natural Resources has a stand on the existence of Chessie or not. At this point, it's a matter of observation."[9]

What mattered most, of course, was not the agency's opinion, but the fact that another, bona fide, sighting had been made in Maryland waters, near Kent Island. The location did not just draw continuity with the Frew sighting; by mid-decade, the media was gradually reimagining Chessie as *Maryland's* monster. That process would be made more difficult if Chessie popped up somewhere on the periphery, or outside, of Maryland's portion of the bay. That earlier sighting in 1984 that Frizzell had mentioned just before the Lohmans came on the scene? By dint of its location alone—near the Potomac River—it was mostly forgotten. The Lohmans, like the Frews, had encountered *their* monster not only near the center of Maryland's side of the Chesapeake but near the center of Maryland too.

A week later, reports of a second sighting near Kent Island came in. Early in the morning of August 9, two brothers from Queenstown, Maryland, Allen and Louis Blunt Jr., were crabbing in the Eastern Bay when Allen saw something large in the water. He thought it was a log and alerted Louis. The creature, which they described as snake- or eel-like and lacking humps, and estimated at about 35 feet long, circled their boat and its immediate environs for an hour. The Blunts periodically spotlighted the creature, causing it to dive and not resurface until the light was switched off; it did not seem dangerous. Allen told the *Bay Times*, "We saw it off and on for about an hour, but as soon as a lot of boats came into the area, it didn't come above water anymore."[10]

In its own terms, the Blunt sighting was nothing special, and it is never included on the list of notable Chessie encounters, but it was among the most important all the same. Here, at long last, was the coveted sighting by an unadulterated, uncompromised, honest-to-goodness waterman—indeed, *two* watermen. In the past, reporters like Burton had boasted of such witnesses, but when their full backgrounds were revealed, they inevitably turned out to be weekend fishermen, pleasure boaters, or tourists, or else their names somehow failed to be disclosed. The Blunts were *actual* watermen who made their living on the water by harvesting what was in it—and they were willing to go on the record about Chessie. It was a milestone in the monster's story, to be sure.

The *Bay Times* certainly recognized the event's significance. "Now seen by watermen," the paper proudly announced in its subhead for the story. Managing editor Lisa Lister, a Kent Island native, was well aware of the skepticism that reigned among the fishing community toward the sightings. "Many local watermen say that on some days a boat can go by and as long as an hour can pass and all of a sudden you may see three or four perfectly aligned bluish-black humps in the water," she wrote in her article on the Blunt encounter. These "humps" invariably turn out to be "just wakes," she explained, but watermen admit they "sometimes appear to be a spiny sea serpent." Instead of yielding to this received wisdom, however, the Blunts became proselytes for Chessie. "You can hear about something being out there and not think much about it," Allen Blunt confessed to Lister, "but when you see it . . . you believe it!"[11]

The Blunt brothers' conversion was a signal moment for Chessie. Watermen could finally come out of the closet and admit to what everybody already knew anyway: that they regularly spotted sea monsters and other strange phenomena while working out on the water and just refused to admit it. Either that, or maybe the watermen who kept quiet did not know the water as well as they thought they did. It was a peculiar inversion of logic that came from the proliferation of suburban attitudes toward the water and lingered just below the surface throughout the rest of Chessie's life span in the public eye. Maybe watermen were *not* the real experts on the water around Kent Island. Maybe work was *not* the main way to know about water and what lived in it. Maybe mere *appreciation* of the water really did confer intimate knowledge of it.

Consider the *Bay Times*' "Island View" feature that ran in the edition reporting on the Blunts. "Island View" interviewed the man on the street about issues in the news, and that week's installment asked half a dozen people if they believed in Chessie. Five out of the six said they did, in one way or another. Three of the believers were convinced that "there's something out there" (two used these very words) but did not specify why. One of them, Bob Dillon of Stevensville, put the focus on the bay. "The Chesapeake Bay is a big place," he explained. "There's no telling what's out there. . . . If I saw it I would believe it." The other two respondents put their trust in science—or something like it, at least. "Being a scientist," Stevensville resident Bob Gabler explained, "I'd say there's a fair possibility it exists." Jim Gick, of Queenstown, did not believe in Chessie as such, but he felt that "the most logical" explanation for the creature was the anaconda theory.[12]

What all these replies have in common, of course, is their reliance on speculation, and even perhaps faith—despite the talk of science. By 1984, Chessie *was* a matter of conviction: one either believed in the creature or not, and the difference between the two camps mainly turned on the question of knowing the water. It is notable that the only dissenter in the group, a onetime waterman from Stevensville named John Hafer, knew the water through labor. "I think it's a figment of the imagination," he told the *Bay Times* flatly. "I worked on the water for three or four years and never saw it and never will."[13] The Blunts' conversion may have made waves, but that did not guarantee that *other* watermen would identify them as humps.

Chessie Central

A week after the Blunt encounter, yet another sighting hit the front pages of local papers. On August 14, the Thomas Creekmore family, from Queenstown, was out boating and went wading in the shallows off the northwest side of Wye Island. At 3:30 p.m., the family of six sighted two humps emerging from the water as the creature swam by, "fairly close to the shoreline."[14] The encounter lasted only 30 seconds, but it was the third near Kent Island in as many weeks, and it caused the *Bay Times* to go into overdrive. When plotted with other sightings on a map of the Chesapeake, the latest report clinched what everybody had already concluded: Chessie had an affinity for Kent Island. The revelation brought the whole Chessie phenomenon to a kind of critical mass. If there had ever been any doubt before, it was now clear that Chessie belonged to the state of Maryland—and especially to the island.

Chessie practically became the *Bay Times*' raison d'être in 1984. The journalistic reasons for this development were obvious. Whatever the Chessie phenomenon was, it seemed to be taking place right in the newspaper's backyard (and thus its domain), and it seemed to be coming to a head. After the Creekmore sighting, the *Bay Times* editorial team openly called for proper scientific study of the monster. There must be *something* to the sightings, the paper argued in a column on August 22, for so many people to come forward in the face of potentially withering public skepticism. "Opinion is divided between believers, non-believers and the open minded. We are in the open-minded group," the paper proclaimed, and "the least we can do—and will do is listen with respect."[15]

A second column in that week's issue, by editor Lisa Lister, admitted to readers that the preponderance of evidence suggested Chessie was more than "a figment of everyone's imagination." The most recent sighting suggested that Chessie "just couldn't resist the idea of being in the paper once more." But, Lister argued, Chessie was not like other unexplained phenomena like Bigfoot or the Loch Ness Monster, which had more than a whiff of tabloid silliness about them. "But there is just something about Chessie that is so believable," she wrote.[16] Lister was less definite about what made the monster so convincing. Perhaps Chessie was just becoming another charming quirk of living in the Chesapeake, an accepted part of the local culture?

Several weeks later, another summer sighting came to light. On October 3, the *Bay Times* printed an account by Grant F. Breining, of York, Pennsylvania, who had been taking his family and a friend to the Coast Guard Reserve graduation on June 26. Breining explained that conditions were perfect during the boat trip from Annapolis to Yorktown: "Our beloved Bay was 'showing off' for our family and guest." With the exception of his son, Breining noted, "everyone on board could be classified as middle age, conservative and prudent boaters. We do not drink when underway. In other words, we were not a frivolous bunch out for a joy ride." As the boat rounded a point, Breining noticed a wake heading toward a marker light. Then a head and humps broke the surface, revealing an "eel-like body." Breining summoned the crew "and we all watched what-ever-it-was swim nonchalantly" toward the light, before it vanished under the water.[17]

The Breining report has been almost completely forgotten in the gallery of Chessie encounters, but it bore all the hallmarks of the very best, and most legitimate, sighting accounts. Its precise navigational information not only invoked the knowledge of experienced sailors but also *evoked* the maritime expertise that had made the Guthrie sighting so compelling back in 1980. Such details lent credence to Breining's descriptions of the creature itself, even though his version of the monster, predictably, fit the mold of the generic Chessie that had become standard since 1982.

Breining's account was convincing enough that it inspired Paul McKnight, the new managing editor of the *Bay Times*, to designate the paper "Chessie Central." Eyewitnesses to Chessie manifestations were encouraged to submit detailed, written accounts similar to Breining's to the newspaper's office, which would act as a clearinghouse for the information. McKnight adopted no position on what Chessie actually was but argued that the mounting evidence justified gathering reports in a single "depository" so they would be easily available for "public use." "The purpose of 'Chessie Central,'" McKnight wrote, "is to simply collect data, screen out obviously false reports and to make available to sincere investigators as much information as possible about the subject of the sightings." Eve Horney, a *Bay Times* staff writer, was appointed to organize the initiative.[18]

In a separate editorial in the same issue, McKnight elaborated on the need for the *Bay Times* to step up. "We thought that," he wrote, "since no one else has done it, and since the *Bay Times* in Stevensville is located in the heart of 'Chessie Country' that, in time, we will be ideal—and enthusiastic—Chessie experts." Musing that the Gloucester sea serpent of the 1810s could have been an antecedent of Chessie, McKnight reflected that the allure of Chessie may derive from the human fascination with the ocean and the sea. Quoting Rachel Carson's view that the sea reminds us of the origins of life itself, McKnight concluded philosophically that Chessie and its ilk prove that "we have not seen nor categorized all life forms; no not yet."[19] McKnight was perhaps channeling the romanticism of Herman Melville more than the scientism of Carson, but both approaches arrived at much the same destination. There was something about the water that was unknowable, mystical, unfathomable—and the solution to that puzzle would be found not in the traditional source of water knowledge, watermen, but in eyewitness accounts of Chessie sightings. *Appreciation*, not *experience*, would reveal the secrets of the Chesapeake.

The *Bay Times* took its new role very seriously, although its sincerity may not have been communicated effectively. In 1985, the paper began distributing coffee mugs with Chessie artwork as thank-you gifts for subscribers. According to a staffer interviewed in the early 1990s, the mug referred to the *Bay Times* as "Your All About Chessie Newspaper" and featured "a drawing of a grinning cartoon serpent wrapped around the letters, 'If You

See Her, Call This Number.'"[20] Sales gimmicks aside, the connections the *Bay Times* drew between Kent Island and Chessie sightings, on the one hand, and between Chessie and the mystical waters of the Chesapeake Bay, on the other, drew together several important threads that stretched all the way back to sightings along the Potomac in 1978. Back then, the Northern Neck of Virginia, where most of the early sightings took place, was just beginning to feel the bite of suburbanization. The Potomac monster had manifested itself as recreational uses of the water crowded out work uses, and as new residents and tourists imposed their urban outlook on the region. The same forces that had been transforming rural Virginia in the 1970s had been at work on the Eastern Shore of Maryland since the 1950s. Kent Island was Maryland's Northern Neck, only at a much more advanced stage in the process. By the 1980s, the only thing that fit in on the island more aptly than a sea monster was a newspaper devoted to reporting its every move.

Waterfront Developments

Kent Island was invaded by a sea serpent once before—but this one never went away. On July 30, 1952, the first span of the William Preston Lane Jr. Memorial Bridge, known informally as the Chesapeake Bay Bridge, was opened to the public. With the bridge's eastern terminus planted irrevocably on the island's western shore, and US Route 50 acting as a conduit, the Pandora's box of post–World War II social transformation spilled out from urban Maryland and Washington, DC, onto the Eastern Shore in the 1950s, loosing suburban development and automobile culture onto an unsuspecting Kent Island and changing it forever. Memoirs of life on the island and the Eastern Shore bulge with accounts of the trauma induced by the arrival of the Bay Bridge. In the mid-1970s, historian Boyd Gibbons concluded that its opening was *the* fundamental moment of change for life on the Eastern Shore—especially the region around Kent Island. "With little difficulty," he wrote, "you can find people on the Eastern Shore who will tell you, without a trace of humor in their voices, that they would gladly blow up the Chesapeake Bay Bridge, if that would return the Shore to its former tranquility."[21]

Yet it is important to remember that what was tranquility for some was paralysis for others. Economically speaking, Kent Island, Queen Anne's County, and indeed the whole of the Eastern Shore were all historically disadvantaged—to a large extent because of their geographic isolation. As historian George Callcott has observed, between 1790 and 1980, "the chief export of the Eastern Shore and southern counties was their ambitious young men—not primarily the poor, or the women, but the enterprising males." These "outward bound migrants were the vigorous young people who were seeking opportunities in the cities and suburbs."[22] For those who remained, good jobs were thin on the ground. "Of those who are employed," Gibbons reported of Queen Anne's County in 1977, "half earn less than $8,200 a year, and one out of four workers makes less than $5,000 annually. There are few well-paying jobs in the county—many are barely at subsistence levels."[23] Agriculture and seafood, the twin industries that traditionally supported the Eastern Shore economy, could only employ so many workers, and both industries were struggling in the 1970s and 1980s. For many Shore residents, even if the Bay Bridge *did* destroy traditional cultural patterns, the cost was an acceptable one purely because new development promised new jobs unconnected with the existing Shore industries. The bridge literally paved the way for new possibilities—for employment, yes, but also for exploring the outside world and admitting new ideas back home.

All the same, it is hard to escape the conclusion that the bridge unleashed something terrible on the Shore, and especially on its island gateway. A choking glut of traffic was the first result.

The year the bridge opened, over a million cars crossed over it. By 1972, that number had reached an estimated 6.5 million. When the bridge's second span opened in 1973, the volume of traffic only increased. By 1980, over 10 million vehicles were crossing the bridge annually. Paralyzing beach-traffic backups became routine on Kent Island by the mid-1960s, bringing Route 50 to a standstill during the summer months and literally dividing the island in half.[24]

The traffic problem precipitated another: roadside development. Residents worried that the Route 50 corridor would become like the Western Shore's Ritchie Highway—a congested string of stores and fast-food joints, culturally detached from life on the island and an exacerbation of the traffic nightmare. Such development had begun within a year of the bridge's opening, when Baltimore land speculator David Nichols opened Kent Island's first shopping center, the Nichols Building, in 1953.[25] A flood of construction followed, and by the 1980s, it had reached a fever pitch, with no fewer than three new strip malls, an outlet center, and countless fast-food restaurants going up in the period between 1982 and 1987. Families who owned property along the corridor eventually resigned themselves to the loss of their land, lifestyle, and birthright due to the overwhelming development pressure, as did Janet Freedman, who wrote about the experience in her 2002 memoir, *Kent Island: The Land That Once Was Eden*. Trudging across her grandmother's farm, now zoned commercial, with a developer, a defeated Freedman reflected, "Standing along the highway, it is doomed to inevitable development."[26]

At the same time that island residents were grappling with the creation of their own version of Ritchie Highway, they were also drowning in a tsunami of residential growth. Between 1973 and 1981, 2,316 acres were developed in Queen Anne's County; 2,297 acres were converted into houses. Between 1970 and 1980, Kent Island's population grew from 3,800 to over 9,000; housing construction increased by 98 percent in the same period. In 1962, almost 95 percent of Kent Island was in agricultural use; by 1982, it was 44 percent. As one gubernatorial candidate warned the Queen Anne's County Chamber of Commerce in 1982, sudden growth was "coming across that bridge just about as fast as it can. That's where the struggle's going to come from. . . . Desire for this kind of life is just going to go boom."[27]

One of the main forces driving suburban development on the island was tourism, which had carefully cultivated an imaginary Eastern Shore ideal. After World War II, and beginning especially in the 1960s, Maryland actively sought to market itself to tourists, focusing on its history, geographical diversity, and water assets. The Chesapeake Bay gathered together most of those themes, so it gradually came to the forefront of the efforts. The state tourism office encouraged the merging of romantic notions of history and the natural world through the mysticism of the bay and the waterman. In 1964, the office's efforts bore fruit when *National Geographic* published an article exploring "Chesapeake Country" that acted as an extended advertisement for the state and bay.[28] Watermen and their Eastern Shore habitat provided the focus for much of the article, but they were presented as a people living in an Edenic paradise characterized by agrarian virtue and a closeness with nature. The physical and financial realities of the seafood industry hardly came into the picture. Sailing and the culture surrounding it was another focus. Author Nathaniel Kenney was himself a sailing enthusiast and had primarily experienced the bay and its watershed through recreation. For him, "Chesapeake Country" was a space shared equally between sophisticated sailors and earthy "baymen" (as he called them), where visitors could contemplate the relationship between humans and nature, and Marylanders and their history, mainly by studying the waterman.

During this same period, publishers began churning out literary works celebrating and exploring the mystique of the bay and its watermen. Books like Varley Lang's *Follow the Water* (1961), William W. Warner's *Beautiful Swimmers: Watermen, Crabs and the Chesapeake Bay* (1976), and *Watermen* (1979), by Randall S. Peffer, laid the foundation for a cottage industry devoted to Chesapeakiana" that persisted through the 1980s and is still at work today. The zenith of this peculiar subgenre was James Michener's epic novel *Chesapeake*. Published in 1978—the same year as Chessie's first appearances—the book relates the story of four families living along the Eastern Shore's Choptank River from colonial times to the 1970s. Michener had made a career out of selecting unique places and writing sweeping sagas about them, and more than any other work, *Chesapeake* was responsible for raising the region's profile and solidifying the tourism narrative of rugged individualist watermen inhabiting what he called a "sequestered paradise."[29] The novel became an instant best seller, and in its wake the waterman and his world gained almost mythical status. It is this subgenre that John Barth riffs on (some would say parodies) in many of his works, especially in *Sabbatical* (1982), where he explicitly conjoins Chessie with Chesapeakiana. "Have we sailed out of James Michener into Jules Verne?" the characters muse, after encountering the monster near the end of the book.[30]

Michener cast an ominous shadow over Kent Island in the 1980s. In January 1982, Queen Anne's County had united with the rest of the Eastern Shore to invest in an umbrella marketing organization called—what else?—Chesapeake Country. They were selling, as one member of the group put it, "quality of life. The Chesapeake Bay is not just a bay, but an area which has a whole feeling associated with it."[31] Maryland tourism boosters found the whole concept so irresistible that they eventually capitalized on it by establishing *two* scenic byways in the mid-Shore: the Chesapeake Country byway in 1998 and the James Michener byway in 2011. When Paul McKnight spoke of "Chessie Country" in 1984, he was not just riffing on a well-known tourism slogan, he was plugging the creature into a mythos pregnant with economic and cultural power. Of the Eastern Shore tourism industry, historian John R. Wennersten has wryly observed, "Weekend mariners can take a cruise on a Chesapeake skipjack or ride on a replica of a Chesapeake steamboat, . . . and hardware stores do a thriving business selling clam rakes and wire crab pots to people who haven't the slightest interest in using them more than once."[32]

Chessie made for a seamless addition to this milieu. True, tourism promoters did not invent the monster, but it still owed much of its DNA, and its cultural cachet, to the industry and its boosters. And they happily exploited Chessie. In October 1984, the Tourism Council of the Upper Chesapeake, which took in all the mid-Shore counties, including Queen Anne's, voted to use the monster as an advertising device. "The symbol of Chessie can be a symbol of the Chesapeake Bay—something mysterious and beautiful," the council's director, Betty Callahan, told the Easton *Star-Democrat*. Callahan's plans for the creature included elementary school poster contests, distribution of flyers, stickers and pins for visitors, and a bus tour for "out-of-town people." The council stressed Chessie's reputed friendliness. It "hopes to attract, not scare, visitors," the paper noted.[33]

Of course, the proximate reason why Kent Island became ground zero for Chessie sightings was the radical transformation of its waterfront from a place of work to a place to recreate. Before the opening of the Bay Bridge, most of the island's shoreline was rural and open for public use, to the extent it was accessible at all. The economy was primarily based on agriculture and fishing—industries that offered little time for leisure. Just as in the Northern Neck, the waterfront was where people worked; enjoyment of it, whether recreational or aesthetic, was a secondary use. The water adjacent to the shore was subject to the same attitude.

With the arrival of outsiders, both landscape and seascape changed rapidly. Developers had been taking bites out of Kent

Island's waterfront since the days of David Nichols, who, alongside his commercial building, had launched a campaign of subdividing waterfront farms for mass residential tract housing in the 1940s and 1950s. By the end of that period, Nichols's firm had opened five major subdivisions on Kent Island, all situated on the bay or its inlets and marketed in the main to affluent suburbanites seeking proximity and exclusive access to water and water leisure.[34] Although Nichols went bankrupt in 1961, the dream of living by the water survived. By the 1980s, waterfront property in the form of condominium complexes either next to marinas or with boat slips attached had replaced free-standing residences as the preferred style of new housing.

The proliferation of condominiums in the 1980s sums up the obsession with seeing the waterfront and playing on the water that underpinned the Chessie phenomenon. One complex, called Oyster Cove, which began construction in November 1985, was designed so that "each unit will be located on the water, offering unspoiled views of the surrounding Wells Cove, Kent Narrows and Marshy Creek, all tributaries of the Chesapeake Bay." Moreover, it had been carefully designed to "combine landplanning with environmental engineering so the development will remain compatible with its vast natural beauty and tidal environment."[35] Even the naming conventions of the new developments were designed to pay homage to the bay. As Wennersten dryly observed of this trend, "Add the word 'Chesapeake' to the word 'estates' and you have a realty winner every time."[36]

In Chesapeakiana, tourism and suburbanization became complementary forces that fed off and strengthened each other, ultimately to the detriment of Kent Island. Visiting there, living there, it was all the same: anyone could be in touch with the Chesapeake merely by being on the island and appreciating the water. Seeing the region's famous sea serpent only heightened the sense of belonging and gave new arrivals, whether they went home to their condos or went home at the end of their vacation, ownership over the water itself. As for the natives who already lived on the island, what did they own? As it turned out, less and less as time went on.

Critical Areas

Residential development of all kinds radically changed Kent Island in a very short span of time. Not only did it displace locals spatially, it also did so economically. Like the Frews with their camcorder, newcomers tended to be more affluent than natives, with occupations on the other side of the Bay Bridge. New arrivals were also culturally and politically very different from the locals, and they simultaneously seemed to want the island to change *and* stay the same, or, in typical NIMBY fashion, at least the same as it had been when they themselves had first arrived. Ultimately, they wanted the island to be "pretty," and they bought waterfront property for its beauty and quaintness. They wanted to tame the wilderness just enough to enjoy its spiritual and social benefits, but not enough to make it less "natural." In a word, they wanted Chessie.

Natives got in the way of those desires, especially when it came to the "messy" lifestyle of watermen. In the years before the mass arrival of outsiders, backyards had been full of the tools of the waterman's trade, junky pickup trucks, bateaux, and working gear. When the tract houses came, the rest of the land had to be kept "pretty," too, and natives ran the risk of upsetting outsiders' sensibilities. Likewise, quaint fishermen on the water were one thing, but when watermen began running their winches just beyond the shore during the early hours of the morning, suddenly they became noise polluters, or they became trespassers, fishing too close to private piers. It did not matter that watermen's livelihoods depended on such "messiness"—outsiders still complained. As island native Louis C. Timms complained in January 1984,

> When the white man settled Kent Island, they [*sic*] didn't run the Indians off with a hodgepodge of houses, 27,000 cars a day, banks, gas stations, fast food shops, condos, shopping centers, malls, and who knows what else is to come and don't forget the bird house-style high rises which are inevitable after all the green spots are used up. Has anyone ever taken time to ask a true native of Kent Island what he would like to see? . . . We could take our families swimming on any beach. We could walk freely around the Bay shore if we so desired. GROWTH, as it is called, put an end to all these good things. These people who have caused the pillage and rape of a once-beautiful part of this state will never earn the right to call themselves Kent Islander.[37]

As tensions increased in the mid-1980s, the *Bay Times* positioned itself as a forum for free debate, and as a unifying force for both natives and newcomers. The paper focused on matters directly affecting its readership and attempted to rally all residents, whatever their backgrounds, to stand up for their right to determine the future of further suburbanization of the island. The *Bay Times* did not oppose development outright, but it did champion watermen and their traditional culture and resisted changes that would put them out of business and out of the community. The stance made strange bedfellows out of two camps normally hostile to each other: natives hoping to preserve the home they had always known, and outsiders who wanted to preserve the charm of the island they originally came to, yet NIMBY away future arrivals, subdivisions, and construction. The paper had always walked this line, though; it had been started back in the 1960s by civic-minded outsiders during the island's first wave of suburbanization. Two decades later, natives Lisa Lister and Eve Horney became successive editors. Both women had strong family ties with the waterman community and advocated for it strongly and unapologetically during their tenures, even as they tried to speak impartially to the concerns of the larger Kent Island community. When Paul McKnight was appointed general editor of the paper and its sister publication, the *Queen Anne's County Record-Observer*, in mid-1983, he took a more forceful (and critical) position on development, fashioning the *Bay Times* into a clearinghouse for information about the explosive kaleidoscope of projects then sweeping the island. It was a model he returned to the next year, when Chessie sightings also exploded in the area.

But the writing was on the wall for native islanders, who simply could not afford to buy waterfront housing. Indeed, the possibility began to emerge that locals would not be able to afford *any* housing at all, as property values rose to meet the higher incomes of newcomers. "It's official: land and housing on Kent Island is now priced beyond the reach of the median income of Queen Anne's County families," the *Bay Times* declared in October 1985. "In plain words most watermen, farmers and other working folks cannot buy a house, rent one or buy even a modest sized lot near the water anymore."[38] In the same way that Chessie elbowed watermen out of their prime status as arbiters of what was in the water, outsiders were edging watermen out of Kent Island life.

Fortunately for watermen, if not all native islanders, the contest for control of the waterfront intersected with a statewide movement afoot in the mid-1980s to preserve land bordering the bay and its tributaries from development. The resultant legislation, the Critical Areas Protection (CAP) Act of 1984, designated any land within 1,000 feet of the bay or its tributaries as a "critical area" on which the health of the Chesapeake depended. The jewel in the crown of bay legislation for many advocates, the law sought to protect wetlands, which absorbed pollutants and furnished habitat for wildlife, and established a 100-foot buffer zone around waterways to prevent development in these vulnerable areas.[39] Although a state commission planned and nominally policed the legislation, responsibility for critical areas effectively devolved to county governments, which were charged with bringing their zoning ordinances into line with the new law.

CAP was a good example of the kind of environmentalism championed by the suburbanites who tended to see Chessie. Although it had the concomitant benefit of helping to save bay wildlife and the waterman heritage connected to it, at its core, CAP was designed to conserve nature for its own sake, and for its aesthetic and consumer value—desiderata common among outsiders.[40] The law was especially controversial on the Eastern Shore because development was really hitting its stride there in the 1980s. Proponents of development—namely, developers and real estate companies—argued the legislation would unfairly deny the economically depressed region the same advantages the Western Shore had already experienced. Even those who normally opposed the development and real estate interests—particularly farmers, whose land increasingly fell prey to those interests—had to agree that there was a whiff of colonialism about the measure: it appeared that the Western Shore, having destroyed its own environment, wanted to ensure it still had a playground left over to use.

For their part, watermen—and Kent Islanders in general—virtually *had* to embrace CAP because the measure offered a solution to rampant development of the waterfront. Watermen, of course, needed to support the legislation because it would ensure their economic future, whereas nonwaterman natives hoped to prevent further destruction of their birthright. Newcomers were motivated by several linked desires: to retain the charm of the island, for which they had paid dearly; to ensure their property investment remained good; and perhaps to adhere to a sense of civic responsibility toward what the region had been—and, of course, many shared the same environmental outlook as the framers of the CAP legislation to begin with. All were probably pleased that the state had taken steps to protect the island (along with the rest of the bayfront), when development had seemed unstoppable.

Although CAP promised much and was written into the Queen Anne's County comprehensive plan that was approved in the late 1980s, in the end it only slowed construction projects on the shoreline and barely touched development on the Route 50 corridor. Partly this was because the new plan had to accept the current distribution of development as a baseline—once built, condominium complexes and tract housing cannot be taken back. As the established "urbanized" area in the county, Kent Island was effectively sacrificed to the monster surging across the bridge.

On the last day of 1986, the *Bay Times* featured a photograph of an oyster whose shell had become attached to a golf ball. Discovered by Warren Sadler, an oysterman with "40 years experience," the bizarre find was made "nowhere near a driving range" or golf course. Sadler concluded that it was the result of "some duffer practicing his swing on the shoreline." For its part, the newspaper noted blithely that "the advent of the Bay Bridge and its accompanying leisure activities has even affected the oyster harvest!"[41] Although the paper passed off the artifact as an amusing curiosity, it captured in one image how completely tourism and development had permeated and transformed life on Kent Island, and how locals had come to accept the new normal.

Chessie was another symbol of the new normal. Like the golf ball, the monster was an alien object imposed on, and grafted onto, Kent Island. It was not a natural outgrowth of the island's culture and history but instead an extrinsic and extraneous thing, brought by outsiders yet nonetheless able to take root there. Of course Kent Island became Chessie Central in 1984—by then, the monster had swallowed the island nearly whole. And the *Bay Times*? It was reporting from deep within the belly of the beast.

Chapter 5

Creating a (State) Monster

MERE DAYS BEFORE SIGHTING REPORTS exploded in August 1984, the *Baltimore Sun* ran a front-page story revisiting the ongoing analysis of the Frew video at the Johns Hopkins Applied Physics Laboratory. Apart from its remarkable timing, the article—which remained disappointingly inconclusive about Chessie's existence—was probably most notable for containing a peculiar appeal. Michael Frizzell of the Enigma Project, the paper reported, was "looking for a state legislator willing to sponsor a bill that will protect Chessie for scientific study."[1] If the need for such a measure were not obvious, Frizzell reminded readers that some sighters had tried to shoot the monster: a bona fide sea serpent would make a neat trophy on a sportsman's wall, or even a useful specimen mounted under glass in a museum.

The appeal paid off. "As a direct result" of the *Sun* article, Frizzell triumphantly reported to fellow cryptozoologists in upstate New York, "political interest in Chessie has been evoked."[2] The source of that interest was George W. Della Jr., a state senator from Baltimore City who had just taken office in January 1983. Della had been following the Chessie saga in the papers and was fascinated by the possibility of the bay having its own Nessie.[3] His interest was not just scientific. By the mid-1980s, the Chesapeake Bay had become the object of a growing and increasingly urgent environmental movement. Alone among his fellow legislators in Annapolis, Della seemed to grasp the monster's potential as a vehicle to address those issues. Chessie was already big news in Maryland: Could the state's own, homegrown water anomaly somehow represent—embody even—the bay and its problems? It was an intriguing hook that the neophyte senator could not resist.

But could the Maryland Senate resist Chessie? That would be the question as the state's legislative session opened in January 1985. The answer, and to some extent Chessie's reputation as a scientific and cultural phenomenon, would depend entirely on the convergence of two otherwise parallel strands of Maryland life that met and merged in Chessie: pop culture and environmentalism. Much as the monster had been conjured into existence by 1970s New Age predilections, its *deployment* was an innovation of the 1980s, a decade when politics was increasingly cynical, culture was commodified, and life was more about being seen than ever before. Chessie seemed to be perfectly suited to the forces shaping its era—but would it be adaptable enough to navigate the more perilous waters of Annapolis?

Been There, Done That, Bought the T-Shirt

Senator Della's "political interest" in Chessie was partially a response to the media's increasing promotion of cryptozoology (and other paranormal matters) in the late 1970s and early

1980s. The television anthology series *In Search Of . . .* (1977–82) set the tone in its first season with episodes covering Bigfoot and the Loch Ness Monster, and other cryptids followed across the remainder of the series' run. Presenter Leonard Nimoy signed off for the last time in the spring of 1982, but reruns lived on. Indeed, two separate Maryland stations showed episodes (unfortunately, we do not know which ones) the Sunday evening before the Frews saw Chessie.[4] *In Search Of . . .* inspired numerous imitators, such as *Ripley's Believe It or Not!*, hosted by Jack Palance and broadcast on ABC from 1982 to 1986, and the Robert Stack–fronted *Unsolved Mysteries*, which debuted on NBC in January 1987 and survived in several different forms until 2002. Neither series was as outré as Nimoy's, but they covered similar ground.

Cryptids and mutant animals, of course, had already invaded cinemas in the 1970s with *Jaws* and *Piranha*, but in the 1980s, their portrayal evolved. Once employed to bludgeon audiences with the theme of "nature gone wrong" or "nature strikes back," now cryptids began to symbolize the joy and wonder of the natural world. Thus, the decade began with *Alligator* (1980), in which a baby alligator is flushed into New York City sewers, grows massive, and terrorizes the city. This was followed by *Splash* (1984), a literal fish-out-of-water story in which Tom Hanks falls in love with mermaid Daryl Hannah, and *Harry and the Hendersons* (1987), in which a Sasquatch comes home to live with a suburban family—with hilarious results. Cryptids could now dramatize both the connection between humans and nature and that between humans and, symbolically, other humans different from them.

One of the more notable films in the cryptid genre was 1985's *Baby: Secret of the Lost Legend*. Produced by Touchstone Pictures, the same Disney subsidiary that had made the incredibly successful *Splash*, the movie starred William Katt and Sean Young as husband-and-wife paleontologists searching the African jungle for evidence of the brontosaurus-like Mokele-Mbembe, with Patrick McGoohan as their adversary. *Baby* was clearly inspired by the expeditions of legendary cryptozoologist Roy Mackal, and it certainly seems to be an attempt to capture the same lightning in a bottle that *Splash* had, in terms of both subject matter and success. Unfortunately, however, *Baby* turned out to be a spectacular flop at the box office. Nonetheless, the movie was notable for bringing cryptozoology to mainstream audiences even as sightings of Chessie and other cryptids were being reported with increasing frequency all over the country, and cryptozoology itself was gaining a higher, and perhaps more legitimate, profile.[5]

Local media outlets probably could hardly believe their luck when it turned out cryptozoology might be relevant in their own backyard. Baltimore-area television stations contributed profoundly to the monster's exposure, with WJZ Channel 13, for instance, producing several newsmagazine segments focusing on Chessie and its eyewitnesses. WBFF Channel 45 connected with Chessie in a more oblique way. The station had been producing a kids' show, *Captain Chesapeake*, since 1971. By chance, the captain, played by legendary TV presenter George Lewis, had a loyal sea serpent companion named Mondy (pronounced "Mone-dee"). Every school day, Captain Chesapeake and Mondy hosted cartoons, told morality tales, and read letters from fans, who could join his club as "crew members." The show was wildly popular among Maryland's kids (and even adults), and Mondy paved the way for Chessie's big splash in local popular culture in the 1980s.[6]

The Mondy connection was just a small part of Maryland's growing obsession with its resident sea serpent. As journalist David Dudley remembered in a 1993 article, "Harry Hughes, Maryland governor back in the early 80s, cracked Chessie jokes. Tourists snapped up cartoon Chessie T-shirts on the Ocean City boardwalk. The now-defunct Chesapeake Restaurant, on Charles

Captain Chesapeake (George Lewis) posing with Mondy the Sea Monster, ca. 1980.
WBFF-TV, BALTIMORE

Street [in Baltimore], opened up a Chessie Room."[7] In 1988, the Annapolis *Capital* reported that the monster had "made an indelible impression on the bay area," with a chain of Anne Arundel County pizza joints and a Chesapeake Beach seafood restaurant all bearing Chessie's name.[8]

What pop culture sensation would be complete without a T-shirt for fans to wear? Featuring a drawing of the serpent in front of the Chesapeake Bay Bridge and a small fishing boat, with a headline above proclaiming, "I Saw Chessie," and the caption "Chesapeake Bay Country," the shirts began popping up around the state in 1983, and they became the most iconic of Chessie cash-ins. Easton resident and businessman Jim Price dreamed up this "cute idea" in 1983, after an interview with Governor Hughes to discuss bay fisheries. Then a member of the governor's Striped Bass Citizens' Committee, Price had gone to Annapolis to discuss funding for a rockfish hatchery, among other subjects. "And he asked me what I thought about Chessie. I was trying to conduct serious business and he wanted to talk about sea serpents," Price recalled with mock indignation. Price took his idea for the T-shirts to local artist Julie Hopkins, who produced the artwork, and his brother, Bill, marketed them. Although Price himself was not a believer, it was clear to him "that Chessie was on the minds of a lot of people."[9] With the T-shirts on people's bodies, too, wherever one looked in the early 1980s, there was a good chance the monster might be there.

Where Chessie really hit the mark, though, was among children. For kids in the 1980s in Maryland, Chessie and Mondy the Sea Monster were portals into a wider world of possibilities that took in not just cryptozoology and weird phenomena but also the rapidly advancing frontiers of science and technology. One of those frontiers was the study of dinosaurs and, in particular, their extinction. The May 6, 1985, issue of *Time* magazine, for instance, introduced the concept of mass extinction events to a mainstream audience in its cover story, "Did Comets Kill the Dinosaurs?" Chessie, with its potential to be a *surviving* prehistorical animal, or an overgrown, out-of-place anaconda, or some other mysterious phenomenon entirely, seemed to localize this wider dialogue and give kids a stake in it.

My brother, Tom, was a great example of the kind of kid whose antennae were always tuned to the media for more information about the paranormal, from UFOs, to Bigfoot, to the Loch Ness Monster. He was a card-carrying Chessie fanatic, complete with one of those famous T-shirts, bought on the Ocean City boardwalk in the summer of 1983. In August 1984, when Tom was 10 years old, the local paper ran a Chessie story using the Guthrie sketch as an illustration. Always a little precocious, he felt moved to write a letter to the editor, offering an explanation for Chessie—at least as depicted in the drawing—

that would have made any professional cryptozoologist proud. Based on information from "my library book," the creature in the sketch, he ventured, resembled "a false-killer whale"—essentially a large, carnivorous dolphin. "My theory has a chance of being right," he asserted, voicing the hopes of monster hunters the world over.[10] For kids like my brother, the idea that a bona fide sea monster might live right in their backyard was just too good to be true. It generated a tremendous amount of interest and goodwill toward Chessie that kept the monster afloat even when sightings dwindled.

Watershed Moments

It is fair to say that George Della's main point of interest in Chessie was the monster's potential to raise awareness about the bay. By the early 1980s, the Chesapeake Bay was in rough shape, and everybody knew it. Finding the correct path to cleaning it up, however, was less certain. Could Chessie make a contribution to that debate? At this early stage, even Della probably did not know, but the connection between the monster and the bay's well-being was a natural one, pregnant with possibilities.

The problem with addressing the bay's health was, at least until the 1970s, mainly one of priorities. Although the Chesapeake Bay watershed includes Maryland, Delaware, Virginia, Pennsylvania, and portions of West Virginia and New York, before the 1960s, it was regulated almost exclusively by Maryland and Virginia as a function of their natural resource management efforts. Both states had well-developed fisheries conservation infrastructures in place on the bay by the beginning of the twentieth century, and as a result conservation drove their interest in bay cleanup. In 1933, policymakers from both states joined with their counterparts in Delaware and Washington, DC, and with federal fisheries officials, to discuss the need for holistic resource and pollution management of the estuary.

They grappled with three persistent problems: pollution, overharvesting, and sedimentation. The meeting had few tangible results, however, and it was not until 1965, when an Army Corps of Engineers study found the bay's water quality to be in steep decline, that serious attention was brought to bear on the waterway's health. It is probably no coincidence that the Chesapeake Bay Foundation, a private environmental advocacy organization, was founded only two years later, in 1967. For the next two decades, its persistent "Save the Bay" campaign helped generate widespread public support for environmentalism among Marylanders and gradually shifted the parameters of debate about bay restoration from *if* to *how*.

The seminal moment in the transition from conservation to environmentalism in the Chesapeake took place in June 1973, when Maryland senator Charles Mathias embarked on a five-day fact-finding tour of the Maryland portion of the watershed. Immediately converted to the cause, Mathias threw his political weight behind legislation that would make the Chesapeake a priority for the newly established Environmental Protection Agency (EPA). In 1976, the EPA performed another water quality study that confirmed the bad news about the bay's health and set the stage for the agency's publication in 1983 of a comprehensive survey and plan of action for the watershed.[11]

For environmentalists at the state level, federal intervention in the Chesapeake was a godsend, not least because the interstate nature of the bay meant that *only* the national government could respond to the crisis in all the complexity it demanded. In Maryland, officials in the 1970s had quickly made state laws compliant with the EPA after its creation, and state political leaders like Governor Marvin Mandel (1969–79) cashed in on the environmental movement of the period with initiatives on clean air and water. But although Maryland was a nationally recognized leader in water and air quality legislation in the 1960s and 1970s, the state restricted its management of the Chesapeake to the conservation of living resources in order to sustain falling seafood harvests.

It was not until the tenure of Governor Harry Hughes (1979–87) that the state moved to ensure that conservation and environmentalism complemented each other. In 1980, Maryland and Virginia established the Chesapeake Bay Commission as an advisory body to legislators at the state and federal levels; Pennsylvania joined in 1985. In 1983, following publication of the EPA report's findings, these same three states, as well as the mayor of Washington, DC, the head of the EPA, and the head of the commission, approved the first Chesapeake Bay Agreement. The agreement expressed a joint desire to save the bay, creating a state and federal partnership for bay restoration and establishing the EPA's Chesapeake Bay Program.[12] A number of other federal agencies contributed to bay research as well, most notably, in Chessie's case, the US Fish and Wildlife Service.

In Annapolis, the legislature passed several bills designed to address pollution and overharvesting in the 1980s, including the 1984 Critical Areas Protection Act and bans on phosphate detergents and rockfish (striped bass) harvesting in 1985. All of these measures, as we have seen with the Critical Areas Protection Act, generated significant controversy, but the Hughes administration took the position that they were for the greater good and that short-term sacrifices would be required in order to ensure the bay's long-term viability. Politically conservative Marylanders, and often even watermen, whose livelihood such laws would help protect, fumed at these expansions of state authority, but there was little they could do to block them in Annapolis, where the legislature was dominated by Democrats and even the *unofficial* consensus was that bay restoration was Maryland's preeminent concern.[13]

In this tense political atmosphere, it is not difficult to imagine policymakers, like George Della, casting around to find just the right hook to sell environmental legislation to the general public. From our vantage point, Chessie seems like an obvious candidate, but in truth, almost nobody thought of the monster in that way before Della came along. In the public imagination, no direct connection existed between Chessie and environmentalism, other than that the monster allegedly lived in the bay. Although Chessie had evolved into a symbol for Maryland and its water heritage, that did not automatically mean that it could perform the same function in a scientific or legal context. But the mystique of the Chesapeake is strong—irresistible even—and a sea serpent in its waters, charged with political potential and beloved by voters across the state, was an opportunity too good to miss. Under the influence of such forces, it was inevitable that the parallel trajectories of Chessie and bay restoration would gradually bend toward each other until they converged.

In fact, the monster had already flirted with environmentalism long before Della came on the scene. As far back as September 1978, when Chessie was more nebulous anomaly than beloved local icon (and not even really Chessie), Associated Press writer Syd Courson authored a satirical piece in which the monster complained about pollution in the Potomac River.[14] In a similar, though more sober, vein, in January 1982, columnist Jack Greer employed the serpent to demonstrate the dangers of dumping dredge spoil off Bloody Point, at the southern tip of Kent Island. The problem was, Greer wrote, "the question of creatures. Although some people have in fact reported seeing something that looks like the Loch Ness monster in the Bay (almost affectionately naming it 'Chessie'), there are other creatures that may favor the trough, creatures of greater economic import."[15]

The most significant (and probably most effective) use of Chessie for environmental messaging in this early period came in a *Baltimore Sun* political cartoon on July 18, 1982. In the cartoon, by Thomas F. Flannery, legendary *Sunpapers* editorial cartoonist from 1957 to 1988, a plesiosaur-esque animal rears up in the foreground, the word "pollution" scrawled on its long neck. It coughs up dead fish and crabs. Typical bay scenery forms the background: on the left, an urban, industrial shoreline (perhaps Baltimore City); on the right, a structure akin to the Chesapeake Bay Bridge; in between, sailboats cruising placid waters under a sky dotted with seagulls and filled with a giant, puffy cloud. A sheet of paper marked "Reduced EPA Standards" floats tailward in the monster's slipstream. The image's caption reads, "Unidentified Creature Reported in Bay Again . . . News Item."[16] The impeccable topicality and sardonic humor were typical of Flannery's oeuvre, and his message was clear: people living around the bay were focusing on the wrong monster. It was an

"Unidentified Creature Reported In Bay Again . . . News Item

Political cartoon by Thomas F. Flannery appearing in the *Baltimore Sun*, July 18, 1982.

observation that Della might have benefited from noting as he launched into his effort to protect Chessie only two years later.

A Specimen Not for Netting

In the latter months of 1984, Della's office coordinated with Michael Frizzell to gather the materials necessary to draft a Chessie bill. In fact, however, they were not setting their sights quite as high as a state law devoted to Chessie's protection. The measure they envisioned, and ultimately drafted, was intended to be a joint resolution. Although resolutions in most cases lack the force of law, in Maryland they are subjected to the same legislative process and designed to be a formal expression of the will of the state assembly. Even if Chessie were not to be protected under an actual law, to receive such an endorsement would nonetheless cement the monster's legitimacy in a very official way.

Della's resolution was modeled on a pair of similar measures that the New York and Vermont legislatures had recently considered (and, in New York's case, passed), calling for designation of the Lake Champlain Monster, popularly known as "Champ," as an endangered animal. Readers may be surprised to discover that the effort to enact legal protections for cryptids is about as old as the environmental movement itself. As early as 1969, Skamania County in Washington State passed an ordinance to protect Bigfoot, a beast as emblematic of the Pacific Northwest as Chessie was of the Chesapeake. In December 1977, four years after the passage of the Endangered Species Act, the US Fish and Wildlife Service issued a press release that wondered rhetorically, "What if they really did find the Loch Ness monster or the legendary Bigfoot of the Pacific Northwest?" The agency's then associate director, Keith Schreiner, conceded that cryptids could be considered for endangered species protection—so long as they were identified scientifically and demonstrated to be under threat.[17]

According to the US Fish and Wildlife Service release, one of the public's premier suggestions for federal protection in 1976 was Champ. At the time perhaps the best-documented sea monster in the world, and *certainly* in the United States, Champ enjoyed a notoriety and a fan base several orders of magnitude greater than Chessie or any other competitor across the nation. Largely due to the efforts of longtime Champ researcher Joseph W. Zarzynski, on October 6, 1980, the Port Henry, New York, Village Board of Trustees resolved to ban visitors who wanted to "harm, harass, or destroy" Champ from the waters falling under town jurisdiction. Zarzynski's concern for Champ was scientific, but Port Henry's was probably rooted more in tourism promotion. At about the same time the resolution was passed, billboards went up proclaiming that the town was "Home of Champ." Nevertheless, in the next few years, the New York and Vermont legislatures followed suit, calling for protection of Champ on scientific, cultural, and even public safety grounds.[18]

For all its growing popularity as a field of science, the cryptozoological community was still a small one in the early 1980s, and it should come as no surprise to discover that Frizzell and Zarzynski were friends and colleagues. Throughout the decade, Frizzell and the Enigma Project kept Zarzynski and his group, the Lake Champlain Phenomena Investigation, updated on Chessie's meanderings, which were regularly printed in Zarzynski's newsletter, *Champ Channels*, and there was a good deal of ongoing cross-pollination and collegiality between the two groups. Due to their close geographical relationship and similar physical descriptions, each monster's existence reinforced the legitimacy of the other, adding weight not only to their believability but also to the credibility of cryptozoology, and to the need for legal protections to ensure the animals could be studied. As early as December 1983, *Champ Channels* reported that Frizzell was "studying the Champ protection resolutions in the hopes of seeing Chessie protected via similar resolutions."[19]

Ironically, Frizzell had actually been beaten to the punch by over two years. Writing in the *Baltimore News American* in April 1981, columnist George Earley had archly suggested that locals ought to take a page from Port Henry's book and get Chessie out of harm's way. "Why shouldn't they [monsters] be afforded the same consideration we give the snail darter, bald eagle, California condor and other officially endangered species?" Earley wondered. "It may be a bit more difficult to put up similar signs at the mouth of the Chesapeake Bay," Earley observed, "but an enterprising town could certainly take Chessie under its wing, thereby advertising itself to natives and tourists alike as a defender of wildlife."[20] Although the column might have been meant as an April Fools' joke, truth was ultimately stranger than fiction.

In the event, as they drafted their measure, Della and Frizzell focused more on the scientific—instead of touristic—value of a protected Chessie. Frizzell was especially concerned that their efforts be taken seriously and not construed as a publicity stunt or lampooned by the media. In early September 1984, he opened up the Enigma Project's files to Della, making available the group's large collection of eyewitness reports for use as testimonials, and he suggested the names of several people he knew would testify in favor of the legislation, including Zarzynski, Bill Burton, and Kenneth King, former secretary of the Maryland Wildlife Federation.

Still, there was a problem. The nub of it was "semantics," as Frizzell told Della, "the rationale, the wording of how to protect an animal not yet proven to exist."[21] It was a point that troubled the senator too. Even with all the public support in the world, if the underlying logic of his resolution failed to hold up, or if its purpose was not clearly defined and justified, his fellow senators would dismiss it out of hand, and probably laugh him out of the statehouse to boot. The easiest solution to this problem would have been to spell out that the monster was really a proxy for genuine bay fauna, and for the bay itself. This was more or less how Champ's boosters had approached their protective legislation, using it as a talking point to generate economic growth.[22]

Laudably, perhaps, Della and Frizzell eschewed the easy road, steering away from using Chessie as a tourism promotion device. But despite their best efforts, that persistent semantic problem made it impossible for them to make a clean break from using the monster as a proxy. The best "scientific" justification for legal protection that Frizzell could muster was that whatever Chessie was, it was probably rare, and if a growing body of reliable evidence suggested it was real, why wait until someone killed it to believe in it? Moreover, if the serpent were rare to begin with, "then Maryland may indeed be gifted with yet another unusual and special feature," and no time should be wasted in making sure that feature was properly protected for the sake of science, and for the sake of Maryland.[23] Frizzell's underlying argument seemed to be that Chessie, like the bay itself, was a precious natural resource that ought to be conserved—a noble sentiment, to be sure, but one that owed as much to tourism as it did to marine biology. It made for a precarious amalgam of environmentalism and Chesapeakiana, and it was hardly the expression of pure scientific inquiry that either man had hoped to project.

By mid-October, Della's staff had drafted a proposed resolution. At his request, its text was essentially the same as that considered by the New York and Vermont legislatures, apart from specific changes made to tailor it to Maryland.[24] The resolution contained a laundry list of recitals outlining the creature's recent history and notoriety, and the reasons why it was worthy of government intervention. At its heart, however, it remained torn between advancing science and perpetuating bay tourism. On the one hand, the resolution posited that discovering—indeed even searching for—Chessie was "a contribution to the fund of human knowledge." It called for recognition by the state of the creature's "possible existence"; its protection against "any willful act" leading to its "death, injury, or harassment"; and encouragement not only of "serious scientific inquiry" but also of reports from eyewitnesses among the general public. On the other hand, the resolution warned that a *lack* of state protection for such creatures might "encourage the use of force or violence that would threaten their welfare and the safety of the people of Maryland and others who enjoy the beauty of the Chesapeake Bay."[25] This latter language was imported directly from the Champ resolutions—a hint, perhaps, that Della had given up trying to parse the logic of Chessie protection. Like Champ, Chessie would have to function as a proxy after all.

Writing to Della on October 28, Frizzell congratulated the senator on the "exquisitely" crafted text, calling it "yet another small victory for the incipient science of Cryptozoology."[26] His confidence was perhaps misplaced. Della's resolution had to convince senators, not scientists. If they did agree to endorse the existence of Chessie, it would be because doing so affected the interests of Maryland, or the bay, not because it would serve as a referendum on cryptozoology. And in any event, the resolution, with its justification rooted strongly in tourism, was manifestly *not* an expression of science for its own sake. If the thrust of the measure had been to enlist Chessie *directly* in support of Chesapeake Bay tourism or environmentalism, it might have been an easier sell. Instead, it forced skeptical lawmakers to choose to believe in Chessie first if they then wanted to believe in the things it could represent. It did not bode well for the resolution, which, now formally submitted and rechristened Joint Resolution 13, was scheduled for review before the Senate Economic and Environmental Affairs Committee on January 29, 1985.

What Della got right about the proposed resolution, at the very least, was the goodwill and great publicity it brought him. A week before it went before committee, the press discovered that Chessie had gone to Annapolis, and suddenly the sea serpent was in the papers again. "Maryland already has a state bird, a state dog, a state fish and a state insect. A state monster may be next," the Easton *Star-Democrat* enthused on January 21.[27] It was a *very* liberal reading of Della's proposal, which never mentioned making Chessie a state *anything*. In retrospect, it is worth asking why Della and Frizzell missed such an obvious trick. In the 1970s and 1980s, Maryland was busy adding to its gallery of state icons. Indeed, in 1985, the skipjack—the Maryland waterman's iconic fishing vessel—was designated the state boat. With water on Marylanders' minds, and a sea serpent in Annapolis, why *not* make Chessie the state monster? Again, it is important to remember that Frizzell, especially, hoped the state would genuinely investigate the monster, not just co-opt it as part of Maryland's identity symbolism. State monsterdom was the *antithesis* of what the resolution was supposed to achieve.

Whatever it lacked in factual accuracy about Senate Joint Resolution 13, the *Star-Democrat* story demonstrates unequivocally that Marylanders really *did* see Chessie as symbolic of the Chesapeake Bay and of Maryland. Despite Frizzell's best efforts, the Chessie the general public wanted to investigate was the one that lived in the realm of imagination. As the *Baltimore Sun* acerbically put it in its daily roundup of odd news stories, Della preferred "to see the creature as a timid mermaid who dwells in the depths of men's minds as well as in the Chesapeake. And that specimen is not for netting."[28]

"I certainly got a lot of press out of it," Della recalled in 2005. Initially, he feared he would be ridiculed—until letters and drawings began arriving in the mail from Maryland children captivated by anything related to Chessie. One letter that found its way to Della's office came from a fourth-grader, Christopher Crowe, of Rockville, who was then studying, as all fourth-graders in Maryland did at the time, Maryland history in his social studies class. "I read your article on 'Chessie' the sea monster of the Chesapeake Bay," he wrote. "I think it is real and naming it the Maryland State Monster is a great idea." Crowe asked for more information and a picture of Chessie, but for good measure, he included at the bottom of his letter his *own* drawing of the creature. "Kids from all over the state were sending me these things," Della remembered, "and I had 'em scotch-taped to the walls. You know, I was having a lot of fun with it."[29]

If I Can't See It, I Can't Vote for It

The Senate Committee for Economic and Environmental Affairs held its public hearing on Senate Joint Resolution 13 on Tuesday, January 29, 1985, at 1:00 p.m., in Room 200 of the Senate Office Building. The agenda was a full one, featuring arguably much weightier affairs than Chessie, including African famine relief and Maryland Critical Areas.[30] Della had invited Robert Frew and Michael Frizzell to testify before the committee. "Bob Frew and I discovered that the Chessie measure . . . had been placed as the last item on an already too crowded agenda," Frizzell later complained. "When the material finally was presented (5 hours later), our proceedings were rushed (by the committee chairman) and the committee members were conspicuously inattentive. In short, they didn't appear to accept the material with the same sincerity or enthusiasm with which it was delivered."[31] "People, you know, were sitting back sort of chuckling about it," remembers Della ruefully. "But I was serious about it."[32]

Unfortunately, no official transcript of the hearing exists, but newspaper accounts provide some sense of the dialogue that took place. After Della introduced the resolution to the committee, Bruce Gilmore, representing the Maryland Department of Natural Resources (DNR), presented his agency's opinion on the legislation. The DNR was considered an "interested party" since environmental matters fell under its jurisdiction. The agency's review in advance of the hearing concluded that it would have no "environmental impact" and that the DNR had no position on whether it ought to be passed. "The Department," it said, "will investigate future sightings and place this species of marine animal on the endangered species list when positively identified by the academic community."[33] It is no great mystery why the DNR's response to the resolution was lukewarm: underfunded, undermanned, and overwhelmed by the bay's health crisis in 1985, the agency no doubt wanted to devote its limited resources to existing problems, not potential or mythological ones. Analyzing and regulating Chessie *after* its formal discovery might fall within the DNR's remit, but it would not be seeking the creature out. As Gilmore testified, "Once he or she is ever seen again and needs protecting, we stand by ready to protect."[34] Practically speaking, the hearing probably ended right there for most of the senators on the committee. If the state agency already charged with managing aquatic natural resources had adopted such an agnostic attitude toward the creature, then what else was there to discuss?

It fell to Frizzell, presenting the Frew film, to make a case for the monster strong enough to pique the committee's interest. Seated some distance from the video monitor, however, members were hard-pressed to identify much in the low-resolution image besides a picturesque shot of the bay, backed with audio of people shouting that *they* were seeing something serpentlike. Unsurprisingly, the senators' reactions were tepid.

"Now I've seen it all," grumbled James Clark, of Howard County, during the showing, when a child on the tape pointed out the extreme length of a putative serpent that he himself could not make out. "You have to be in a certain state before you see this thing," he quipped to reporters afterward. Committee chairman Norman Stone, of Baltimore County, was a little more diplomatic: "I saw the shoreline. I'm not sure what else I saw."[35]

The committee's blasé attitude exasperated Frizzell. For him, the Frew tape constituted positive proof of the serpent's existence, or at least of an anomaly that had proved it was worth investigating. Its dismissal out of hand, and the short shrift Della's resolution seemed to receive in general, left him feeling frustrated. "You got these guys crowded in this committee room, looking at this little, you know, tiny television monitor," he recalled with chagrin in 2005. "And they're all fifteen, twenty feet away from it. Of course they're not going to see it. . . . And nobody made an attempt to get up from their chair and move closer. It was like, okay, they're looking at this black and white monitor for a few seconds, and then, 'Okay, what's next on the agenda?'"[36]

Witness testimony also failed to move the committee, although it raised some interesting issues that neither Della nor Frizzell seem to have thought about when they were composing the resolution. Commenting on his video, for instance, Frew stressed the monster's putatively benign nature. "It swam by kids and none of them is missing," he testified, "so it doesn't seem to eat children."[37] The offhand comment, probably meant half-jokingly, would backfire in short order, as we shall see in a moment. But Frew's assertion that Chessie was harmless, and that, by implication, it therefore deserved legal protection, requires some explication.

One of the curious through lines of Chessie's biography is a consensus, from virtually the first sightings in 1978, that it—whatever "it" was—was "friendly." Witness after witness, from the Potomac to Kent Island, described Chessie in almost affectionate terms. Ann Kyker, for instance, in one of her few statements on that first sighting, told the *Richmond Times-Dispatch*, "I don't think it was a scary type thing."[38] Her neighbors, Howard and Myrtle Smoot, were less sanguine, however, and it was Howard Smoot who famously shot at the creature—an act that ultimately helped justify Della's proposed resolution. It was actually the Smoot encounter that led to the monster's widespread reputation for friendliness. According to newspaper reports, at a meeting of the bistate Potomac River Fisheries Commission on August 18, 1978, chair and Virginia Marine Resources commissioner James E. Douglas apparently intended to ask for a resolution declaring the Potomac creature a "friendly monster" in order to protect it from harassment. If this had taken place, it would have anticipated Della's action by six years, but no record of such a commission resolution, nor even official discussion of the creature, exists.[39] Perhaps the subject of a monster in the river was just deemed too frivolous to take up formally, especially when issues that immediately and significantly affected watermen's livelihoods were at stake.

Still, Chessie's congenial reputation—backed by virtually no proof whatsoever—became one of its abiding traits, even more durable and recognizable than its physical description. The theme really took hold when Chessie crossed into Maryland in 1980. Trudy Guthrie set the tone when she admonished the papers after her encounter not to "bill this thing as a monster." When it printed its interview with Guthrie, the *Richmond Times-Dispatch* went one step further, asserting that "there is general agreement that Chessie is friendly, but somewhat shy."[40] In truth, no such consensus existed, except perhaps in the media echo chamber, but it became a familiar refrain for the rest of the decade. In 1982, for instance, the Frews were said to regard Chessie as an "adopted pet"; a couple of years later the creature was described as "shy, but not threatening in any way."[41]

To the extent that eyewitnesses had not been harmed by the serpent, such claims were superficially true. They would pay off in the early 1990s when the mythological Chessie merged with an *actual* manatee—a friendly sea creature if ever there was one—swimming up and down the Atlantic seaboard. But in 1985, Chessie the sea serpent was still a big unknown; no one really knew what the creature even *was*, let alone whether it was dangerous. It was not even clear that Chessie was a single animal, or an animal at all, or even the same object being sighted on each separate occasion. Frew's testimony, well meaning though it was, was based purely in wishful thinking, not evidence. No doubt the senators, many of whom were no-nonsense businesspeople as well as experienced and skeptical lawmakers, had seen—and resisted—plenty of that in their careers by the time Chessie swam into their lives.

Frizzell's own testimony, a written copy of which survives, focused on the scientific importance of Chessie and the preponderance of evidence that seemed to validate its presence in the Chesapeake. "If the existence of an apparently rare animal is documented by solid and credible testimony as well as photographs, videotapes, and other forms of substantiation," he testified, "should we delay in protecting that animal because an actual physical specimen has not been caught or killed?" Moreover, he argued, the bay "is one of Maryland's most valuable natural resources"; no one would attempt to make the case that its known "rare or endangered" wildlife was not in need of special protection. It was only because of Chessie's murky scientific status that it was not a shoo-in to receive similar status.[42]

Reminding the committee that well-documented cryptids had been protected under legislation elsewhere, and that science regularly identified new species, Frizzell explained that it was "conceivable" that the bay could be the site of the next such discovery. But what if Chessie were killed before that could happen? Frizzell described a scenario in British Columbia, where a $1 million, dead-or-alive bounty had been posted for proof of that region's Lake Okanagan monster. "With stakes that high some people might resort to rather dangerous extremes to claim such a prize. What a shame it would be, if the creature they are hunting is only one of five left on earth. I sincerely hope," he concluded ominously, "that a similar situation never develops concerning the Chesapeake Bay."[43]

A letter of support from Joseph Zarzynski addressed issues similar to those Frizzell raised. The Enigma Project's efforts, Zarzynski wrote, had elevated the monster "from the category alongside the tooth fairy into the respectable arena of cryptozoology." Chessie was "possibly the coelacanth of the late 20th century"—here, he referenced the so-called living fossil fish rediscovered near Madagascar in 1938—that deserved "immediate environmental and scientific attention." "Have we inherited this planet and its residents from past generations?" Zarzynski asked the senators. "No! Rather we have temporarily borrowed it from our children and their offspring. What a grand and very sensible gift the Chessie resolution will be for them, and also serve as a blueprint for biological diversity and a symbol that Mankind must share this Planet with the rest of the living organisms."[44]

Worthy as their commentary was, Frizzell and Zarzynski were probably being too high-minded. Mid-1980s environmentalism in Maryland was less about ideology or high concepts and more about practical management and decision-making. Livelihoods were at stake. Symbolic gestures and grandiloquence could scarcely solve the problems facing the bay, no matter how well intentioned they were. We can imagine committee members clutching their heads in frustration at the disconnect between what they were hearing and seeing and the real-life implications of passing another bay-related moratorium during such tense times.

And surprising as it may seem, the committee took the environmental implications of a large, unidentified animal in the bay very seriously indeed. Noting that the resolution's projected budget requirement was zero—usually a fanciful number in government bureaucracy—Senator Sidney Kramer, of Montgomery County, wanted to know how the state could "protect it and feed it without cost." That question led him to another: "What does it eat?" It was an important and incisive question . . . that no one could really answer. Robert Frew had inadvertently raised the issue during his testimony, but his answer had been a throwaway quip. Della now essayed his own, responding, "Not you sir, or me."[45] These jests were not exactly high comedy, but they *should* have left the committee laughing indulgently at least.

Instead, the question of Chessie's eating habits actually seemed to sour the senators toward Della's resolution even further. James Clark, already the most openly hostile of the bunch, had a clear, if derisive, answer to the question: "Rockfish." And if Chessie ate rockfish, Chessie was bad news. Due to its abundance in Maryland, the fish had been made a state symbol in 1965. Now, 20 years later and 28 days before this very committee hearing, on January 1, 1985, the fish, "whose name has become synonymous with both the bounty and decline of the Chesapeake Bay," as one paper put it, became "banned from the nets, hooks and dinner plates of Marylanders." The controversial moratorium lasted through the end of the decade and was a persistent source of heated debate. If Chessie fed on rockfish, as Clark concluded with impeccable logic, "that would make it unlawful."[46]

Before we chastise Clark for small-mindedness and prejudice, it is worth remembering that the main theory explaining Chessie's intermittent appearances over the years was related to the rise and fall of the resident *bluefish* population. Although Clark wore his skepticism on his sleeve, the concern he expressed was both legitimate and topical. What if Chessie explained why *rockfish* populations were so low? What other populations was it affecting? What if it were part of the *problem* in the bay, rather than something to be preserved? As silly as it may seem, the question marks that hung over Chessie's diet may well have been the clincher for the senators that day. Chairman Stone summed up these worries when he told reporters, "I'm not being facetious when I say if we find out what it is, we may really not want to protect it. We may want to get rid of it."[47]

News of the committee's tepid response evoked mixed opinions among the public. "For shame," reproved Tom White of the *Baltimore News American*, a few days after the hearing. "Chessie should be as much a part of Maryland's established animal life as the snallygaster."[48] Others were not so enthusiastic. Out of eight people interviewed for the Annapolis *Capital*'s "You Said It" column published February 6, six thought the idea of protecting Chessie was either preposterous or a waste of taxpayers' money. The remaining two interviewees showed only lukewarm support.[49] Even Della himself seemed pessimistic about the resolution's passage. Interviewed by a St. Louis, Missouri, newspaper a couple of days before the resolution was to be voted on, the senator confessed that he was "not sure the committee will take his resolution seriously enough to approve it."[50]

Ultimately, Senate Joint Resolution 13 never left committee, failing by a vote of seven to four on February 6.[51] As Prince George's County senator Arthur Dorman tersely explained, "I can't see it, so I can't vote for it."[52] Looking back, neither Della nor Frizzell felt that the committee took the measure seriously. "My colleagues at the time," remembers Della, "I mean they were real staid, stiff people. And they were sort of, I suppose, chuckling at me."[53] Frizzell agrees, but with perhaps greater bitterness: they were "just too apathetic to consider the possibility that there might be something strange migrating through the Bay from time to time, or living there, or what have you. It was . . . completely off their radar of interest."[54] Recalling the episode in 1993, Bill Burton was more succinct and less charitable: "Our stupid legislature just laughed it off," he griped, still vexed almost a decade after the fact.[55]

The resolution's failure provoked a minimal response from the public. "Chessie monster gets no respect," the Annapolis *Capital* lamented the day after the vote. "If a giant serpent-like creature known as Chessie really prowls the Chesapeake Bay, it better watch out," wrote the paper.[56] The sharpest criticism came from Wilton, New York, in a letter from Joseph Zarzynski to the editor of the *Baltimore Sun*. The committee's refusal to acknowledge that Chessie might exist, Zarzynski wrote, was "shocking" and reflected a lack of scientific curiosity in Annapolis. Especially receiving his contempt was Senator Dorman's reductive, "seeing is believing" attitude. "Hey, I haven't seen God," Zarzynski countered, "but I believe he exists." He concluded by expressing hope that Chessie might yet be protected by a municipality, as had been done for Champ in New York.[57] Alas, it was not to be.

As it turned out, the *Sun*'s prediction had been right: Senator George Della's fishing expedition returned empty-handed. Senate Joint Resolution 13 received too few votes to make it out of committee, and the effort to protect Chessie in an official way sank without a trace. The defeat was certainly no political impediment for Della, who went on to serve in the Maryland Senate until his retirement in January 2011. In retrospect, however, the Chessie phenomenon was significantly damaged by the experience. In a way, the hearing before the Economic and Environmental Affairs Committee, which had the authority to confer official status on the monster but did not, acted as a kind of referendum on Chessie. Although sightings would still come in, and the creature would still retain much of its mystique, the stir surrounding it had been dealt a subtle—but ultimately mortal—blow. It was not obvious at the time, but Chessie had changed course and was en route to becoming more of a curiosity than a sensation.

Chapter 6

A Chesapeake Bay Story

THE MARYLAND SENATE'S REJECTION got Chessie off to a bad start in 1985, and the rest of the year saw little improvement in the monster's fortunes. Sightings were few and sub-optimal, and the "official" blow to the creature's scientific credibility gave the media an excuse to go back to covering it with less sobriety. But much as it might have seemed that way to the Chessie faithful at the time, Chessie did not go to Annapolis only to become extinct. In fact, the next step in the monster's evolution was about to take place, and it would result in another heyday for the creature.

As it turned out, only a few miles up the road from the statehouse there was a man who was waiting for exactly the kind of opportunity Chessie represented by the mid-1980s. The man in question was Glenn Kinser, and at the time Chessie came before the Maryland Senate, he was working as the field supervisor for the Annapolis Office of Ecological Services of the US Fish and

Wildlife Service (FWS). Much like George Della, Kinser was looking for a hook on which to hang new outreach efforts, especially in elementary education, that his office was developing specifically for the Chesapeake watershed. Kinser was a bit too scientifically minded to take legal protection for Chessie seriously. As he saw it, the gesture was mostly symbolic and would have complicated rather than clarified bay restoration efforts if it accomplished anything at all.

But Kinser's interest was still piqued by the conceit that the monster was just another threatened Chesapeake animal—if a very exotic one. Could the mythical Chessie be lassoed and enlisted to help the creature's very real, and also very threatened, fellow wildlife? It was a replay of Della's Chessie-as-proxy strategy, for sure, but perhaps handled more plainly and with less earnestness this time around. This version of Chessie would be a proud surrogate, a conduit through which kids could encounter both the wonders and the problems of the bay. "I thought," Kinser recalled in 2005, "maybe if I had this monster introducing all his friends, we can find that the kids are going to be interested in the monster, and then they're going to be interested in the other species that Chessie is highly recommending."[1]

In one way, Kinser's vision was a demotion for the monster, pushing it to the sideline of its own story, but it also capitalized on the years Chessie had spent gradually accumulating meaning and symbolic power in Maryland and the Chesapeake. Could Kinser successfully draw together all of the threads of the monster's career from across the previous decade and finally make them serve a greater, more unified purpose? And was Chessie a sturdy enough icon to carry such a heavy burden? Kinser was about to take probably the greatest risk of his *own* career in order to find out.

Animal Magnetism

The path Kinser took to Annapolis was nearly as serpentine as Chessie. Born in Ohio, he graduated from Kent State with a biology degree in 1962. He received his PhD in ecology from Indiana University in 1967, after which he taught at small colleges in New Jersey and Florida. The FWS hired Kinser in 1971, and in 1977 he was promoted to field supervisor of the Annapolis Ecological Services Office, a position he held until 1988. Under Kinser's leadership, the Annapolis office piloted the Chesapeake Bay Estuary Program (CBEP), which was spun off into its own, independent office in 1988, with Kinser serving as its supervisor until 1992, when he moved elsewhere within the FWS. At his death in 2013, colleagues remembered him "as an excellent scientist, independent thinker and innovator, mentor to many Service employees, and a great humorist."[2] Kinser would bring all of these qualities to bear on his engagement with Chessie.

The serpent's ignominious encounter with the Maryland Senate took place at the midpoint of Kinser's 15 years in Annapolis, right when over a decade's worth of environmental advocacy was about to come to fruition in the form of the CBEP. The program, in development by then for a few years, was Kinser's answer to deficiencies he had identified in the FWS's environmental advocacy process. For instance, the Ecological Services offices, in Annapolis and more generally, were limited in their authority mainly to granting or denying permits to developers. With responsibility for administering large territories with only a small staff and an even smaller budget, the offices were perpetually overwhelmed by the number of projects up for consideration and could at best pick and choose their battles. Even then, the offices' veto power was poor, at once both a feeble safeguard against large-scale development and a blunt object incapable of responding to the varied and nuanced causes of environmental problems. For Kinser, it was clear that unless the FWS took a more proactive, preventive role in the bay's restoration, its health would just keep going downhill.[3]

Early 1985 was a propitious time for the Annapolis office to innovate. With the help of Senator Charles Mathias, whose 1973 tour of the bay had initiated the federal restoration efforts Kinser was now carrying out, the new CBEP received $1.5 million in federal funding for fiscal years 1986 and 1987. "I was on the Chesapeake," Kinser reflected with characteristic modesty, "and that was the new hotspot."[4] The program would concentrate its efforts (and funds) in three broad areas: preservation of living resources, limiting water pollution, and educating residents of the watershed about the impact of their actions on the bay's health. The last goal was especially crucial for Kinser. There were signs at mid-decade that the public's commitment to long-term, sustained environmental initiatives was waning, and FWS officials also worried that individuals living in the watershed failed to appreciate how profoundly their actions could affect the environment. Kinser thought that better education in the present could allay future bay problems.

What Kinser was looking for was a way to interest *kids* in issues relating to the bay. His office already produced traveling presentations, poster campaigns, and reams of informational pamphlets—all directed at adults. But if the CBEP had any hope of effecting change in the long term, it would have to reach out to children, instilling in them a conservation ethic. To reach that target audience, Kinser knew he needed something more alluring than government reports—and that is where Chessie entered the picture. Recalling the hearing on Della's resolution, Kinser invited Michael Frizzell to come present his findings on the monster. He did not leave the meeting believing in Chessie, but he was at least convinced that the serpent could do the job he had in mind. "I thought, well, it's something you can use for education," Kinser later recalled, and "something that would make people interested in the bay."[5] With the extra funding now available to him, Kinser embarked on a project to create a publication—potentially a coloring book—specifically designed to teach kids about the Chesapeake, with Chessie as its spokesman. "What better than monsters," he recalled in 2005, to get children excited about a subject?[6]

Skeptics might insist that the Chesapeake was literally overflowing with icons more suitable than a mythical sea serpent to serve as its cheerleader. As we have seen, the region—Maryland in particular—had, by the mid-1980s, built a whole industry out of promoting Chesapeakiana. Any number of symbols could have been cast in Chessie's role, from the skipjack, to the Chesapeake Bay retriever, to the oyster, rockfish, or blue crab. Kinser himself proposed the canvasback duck for the job, since the animal was both locally recognizable and notoriously threatened by habitat destruction. But, he later admitted, "people did not seem to relate to it as well." The blue crab and the rockfish also garnered some interest, but neither had the right "sense of allure" or "mystique."[7] The monster's selection to represent the bay was not without controversy at the FWS, as we shall see, but it reflected the kind of lateral thinking that Kinser was known for in the service. The virtue of Chessie was that it was *not* real: it was a "third party," as Kinser put it, that "could talk about all the rest of the things that were important that we really wanted to get at. . . . It stood aside from the canvasback or whatever. It stood aside from the things we were using on a daily basis."[8]

Today, environmental scientists and scholars might label Chessie, at least as Kinser envisioned the monster, an example of "charismatic megafauna"—that is, an animal that symbolizes national or regional character (like the bald eagle) or attracts public attention to a particular environmental concern (like whales).[9] Chessie had been headed in that direction for a while, and its flirtation with legal protection had solidified the connection, even in failure (and perhaps somewhat against the wishes of Della and Frizzell). Harsher critics might even accuse Kinser of encouraging the so-called Bambi effect, by which the public comes to sympathize with animals perceived as "cute" or "lovable" (like the eponymous deer), at the expense of less attractive, but potentially more endangered, wildlife.

Far from rejecting such criticisms, Kinser embraced them unapologetically. They were the *exact* reasons why he believed the monster could be so effective. Chessie had already been established in the public mind as a friendly monster and a surefire draw for kids—those were its major selling points. If it was lovable enough to generate public interest in the issues facing the bay, then that was worth fetishizing the monster and engaging in some public relations sleight of hand. The monster would fulfill what had perhaps been its destiny all along: as Kinser put it in 2005, "it will be a friendly vehicle to advertise what it is that we're talking about."[10]

Nor would Chessie be the first "character" employed by the US government to advocate for nature. The monster's closest antecedent was Woodsy Owl, the US Forest Service's anti-litter campaign spokesman. Introduced in 1971, Woodsy—with its distinctive costume and motto, "Give a hoot, don't pollute!"—was designed especially to appeal to kids. In spite of early opposition from Interior Secretary Rogers C. B. Morton (who ironically hailed from the Eastern Shore of Maryland), the character gradually developed into an environmental icon in its own right and provided an alluring model for Chessie in the later 1980s.[11] In turn, both Chessie and Woodsy owed whatever cachet *they* might have had to the granddaddy of all charismatic megafauna, Smokey Bear, whom the Forest Service began using in 1944 in an information campaign to prevent wildfires.

As it happened, there was rather little else to say at that moment about Chessie as a zoological matter, anyway. After the heights of 1984, sightings were poor in the new year, and Kinser and his staff might be forgiven for thinking the creature had run its course as an "actual" phenomenon and was available to be repurposed for their own ends. The only major sighting in Maryland in 1985 took place near Kent Island, on March 4. Local restaurant proprietor Dave Harper and his daughter, Susan, residents in the island's Cove Creek development overlooking the Eastern Bay, were alerted by their dogs to two mysterious objects out in the water. One appeared to be snakelike, about 9 inches in diameter and 10 feet long. The body of the other was obscured, but each had a head "the size of a large duck." After watching the objects for about a minute, Harper went to get his camera, but—in a now-familiar story—by the time he returned, the objects were gone. "I thought I was seeing an optical illusion, I didn't even think of Chessie," Harper told a reporter, perhaps saying more than he realized. "If I am lying, may my business and home burn to hell."[12]

Fortunately, such extremes were not necessary; the *Kent Island Bay Times* was more than willing to believe, and print, the story, especially when the monster made a return appearance two days later. This time, a workman at the Harpers' house, Mike Clark, made the sighting. Clark, for once, had the presence of mind to take some pictures, but alas—again, in familiar fashion—they turned out poorly.[13] Reporter Eve Horney took the Harper and Clark accounts at face value, and there is no reason why we should not too, but it is worth noting that before the year was out, Harper's restaurant, Hemingway's, which overlooked the bay close to the eastern terminus of the Bay Bridge, was actively employing Chessie and Harper's encounter as an advertising hook.[14]

Although Harper's sighting made a local splash, news editors outside the island mostly passed on the story—an indication that Chessie's star was receding, especially in the wake of the "state monster" fiasco in January. Still, Bill Burton attempted to prop up the monster with one of his periodic updates on its status in the *Sunpapers* in May 1985. The article focused ostensibly on an April 5 encounter on Baltimore County's Back River, but

Burton's real concern was the ongoing lack of any photographic evidence of the monster since the Frew tape. This time, Nancy Gabriszeski had watched the creature slither by her waterfront sunporch for five minutes but had been so transfixed by it that she had failed to grab her nearby camera and snap a picture. "What is needed," a frustrated Burton lamented, "is one photo of Chessie clear enough that skeptics will be convinced she really exists."[15]

But with fewer sightings and virtually zero evidence, the silence—if not the voices of the skeptics—was beginning to drown out the believers. The only other sighting that went public in 1985 took place in the Northern Neck, Chessie's home waters. Wading in the shallows of the Potomac while fishing, Lydia Bowles saw a creature that fit the monster's typical description, only *this* creature seemed to be much larger. The *Richmond Times-Dispatch* wondered if Chessie had grown, or perhaps started a family.[16] Whether the monster had increased in size or number was of little consequence, however, as new reports remained few, a fact that the paper starkly pointed out.

Columnist Charles McDowell, the *Times-Dispatch*'s resident wit, could not let Bowles's encounter go by without lampooning it. McDowell offered some "guidelines" to improve potential future sightings, by way of pointing out how parochial and picayune the Chessie phenomenon was. "Don't be intimidated by the scoffers' notion that science rejects Chessie," McDowell observed, in a tone that suggested that scoffing was *exactly* what people ought to do when confronted with the possibility of the monster. Although McDowell had voiced such skepticism in the past, it had been in the face of numerous sighting reports. Now, with only a handful of mediocre sightings to suggest the monster really existed, the science seemed definitively to be on his side. It was a bad omen for Chessie, at a time when the creature really could have used some good press.[17]

In this context, one might wonder why Kinser wanted to hitch his wagon to Chessie's star, or, indeed, whether Chessie was even a viable symbol at all by mid-1985. The truth of the matter was that Kinser had zero interest in Chessie for its own sake. As a trained biologist, he did not believe in the monster—or at least that it was a "monster" and not a misidentified animal or object, or an optical illusion—and so whether Chessie was "real" had no bearing at all on its use as a spokesman for the environment. If anything, the paucity of sightings and lack of photographs *added* to Chessie's mystique, opening up even more space in the public consciousness for the *imagined* Chessie and clearing the way for the Annapolis office to shape the monster in a way that would be most convenient to its ends. The monster they created would go on to become the face of Chessie for the next decade.

Build a Better Sea Monster

Kinser envisioned his office's Chessie publication as a coloring book in an attractive and readable style similar to the famous Dr. Seuss books. "So we were going to hit two big demographics," Kinser recalled, "the parents who might sit down and like this Dr. Seuss verse and the neat clean pictures that were with it, and also their children that they would be reading to."[18] In the past, the FWS and other conservation agencies had published environmentally themed coloring books, but Kinser felt they had been too sterile and teacherly. With the new federal funding now available to him for fiscal year 1986, Kinser hired several new staff members, including intern Jamie Harms. Harms was assigned to comb through stacks of educational materials marketed by environmental agencies to children, in search of a useful template or model for the Chessie book. The closest samples to what Kinser had in mind were *Maryland's Natural Resources*

Coloring Book, produced by the state's Department of Natural Resources, and two FWS titles, the *Wetlands Coloring Book* and *Nature Series: Fisheries and Me*. Kinser, Harms, and other staff evaluated these books in detail, commenting on the quality of the art, the content and style of the text, and the "colorability" of the line drawings used for illustrations. Kinser found them all to be clunky, out of touch, and poorly drawn—if anything, they were examples of how *not* to produce a children's reader.

Some of the staff's written comments on these earlier publications were hilariously pointed. The *Wetlands Coloring Book* came to the Annapolis office with an impeccable pedigree, having been illustrated by Jack Elrod, author and artist of the famous conservation-minded newspaper comic *Mark Trail*. Alas, the connection was not enough to save the book from the staff's disapproval. In one illustration, a man, a boy, and a girl canoe down a river. The boy trails his hand in the water. A duck flies into the scene, which is set against a wide river, with marsh and woods in the background. The man tells the children, "Wetlands are just that—Wet Lands! This coloring book will help you learn more about wetlands and the animals that live in them. . . . Let's get started!" On one copy of the page, "BORING!" is scrawled across the bottom. Another comment noted that the canoe scene was "too crowded for crayon coloring," while the duck came in for criticism because it was "larger than boy in boat." Not even the boy in the boat was safe: next to his hand dangling in the water the commentator joked, "hand destroyed by piranhas"! As the reviewer dryly concluded, "Nobody has been particularly excited about that book."[19]

Fisheries and Me was drawn in a more child-friendly style, closer to what Kinser had in mind, but it still received harsh, if comically put, criticism. Again, one scene was singled out. It depicted a miserable-looking salmon swimming in the foreground, a pile of fish eggs next to it, with a tiny fishing boat in the background. Some scribbled trees and a lighthouse completed the picture. At the base of the page, a caption read, "Rivers, streams, ponds, and lakes are important in the early and late stages of the life cycle of salmon. Oceans provide habitat in the middle stages." The book, a reviewer wrote, was "not as somber as the *Mark Trail* book, but it also has some problems, like the expression on the fish's face and that clump of dotted things (they must be eggs, but I'm not sure many kids would figure that out) and the rather abstract shapes that are trees maybe? The lines should be a bit thicker for coloring." Other notes pointed out that the sailboat was too small for coloring, that the page caption was "terrible," and that "[in] general fish and birds probably shouldn't have these kinds of expressions!"[20]

Armed, then, with a pretty clear idea of what they *did not* want their coloring book to be like, Kinser's staff drew up specifications for their own project. Their guiding principle was that the book's artwork was the key to its success. It had to be just right—visually appealing yet capable of conveying complex, if simplified, ideas in compelling ways. The broad guidelines given to artists were simple: the illustrations could not be too intricate, so that children would be able to color them, and they also had to "work in [the] theme of [the] conservation ethic."[21]

The most crucial component of the illustrations, naturally, was the design of Chessie, which would inform the rest of the artwork in the book. Perhaps surprisingly, there was a genuine attempt to depict the monster "accurately," so much so that a short description of the creature—the classic "telephone pole with humps on its back"—was included in the guidelines Harms wrote for prospective artists. "There are no good pictures of it," the instructions explained, "so use your imagination, but don't stray completely into pure fantasy." One fantasy Harms specifically told artists to avoid, in spite of many eyewitness descriptions, was the monster's reputed "vertical undulation" method of propulsion. "Chessie presumably swims side to side, like a snake, rather than with an up and down motion," the guidelines noted with an air of finality. Scientific accuracy mattered even more when it came to depictions of *actual* bay animals. It was not necessary for wildlife to be anatomically correct in the pictures, but they did need to be "recognizable." Readers "should be able to distinguish a blue crab from any other kind of crab," Harms advised.[22]

In the event, six artists submitted sample art, and office staff judged them. Two of the unsuccessful entries have survived to provide a glimpse at the Chessies that might have been. One, by Marianne Lydecker of Kensington, Maryland, bore a resemblance to Casper the Friendly Ghost. It was, perhaps, a bit *too* simple and childlike in style. FWS staff member Charlie Rewa, by contrast, drew Chessie as a cartoon dragon, complete with a puff of smoke exiting his nostril in one picture.[23] Harms praised Rewa's submission as "among the most imaginative of the bunch," but she still vetoed it, no doubt because it imagined a Chessie that was more threatening than threatened.[24]

The successful submission belonged to Dave Folker, the teenage son of an FWS staffer. Folker's Chessie had a distinctive look that not only found the happy medium between sweet and scary, it also mimicked the style of cartoon characters kids might see on Saturday-morning or after-school television. There was an especially strong hint of Scooby-Doo in Folker's Chessie design. "There was just so much difference between his pictures, which were nice and clear," Kinser remembered in 2005, "and these others which tended to be very intricate, and very fine, and very laborious."[25] Reviewers within the FWS agreed, describing Folker's art as "clever," "engaging," and "appropriate for the 4-to-10 age target audience with its mix of simple and more complex illustrations."[26]

Keep the Focus on the Bay

The book's text experienced a more difficult gestation than its artwork. Harms's initial plan established two main goals: on the one hand, to introduce children to the bay's endangered wildlife, and, on the other, to broach the subject of pollution. Harms envisioned the coloring book as a sort of travelogue, with Chessie first visiting fellow wildlife and then pointing out environmental problems. The featured wildlife, including the rockfish, blue crab, great blue heron, and canvasback duck, would be drawn from the FWS's recently developed list of threatened animals, called "Species of Special Emphasis." Conveniently, when "personalized" for use in the Chesapeake, the list's name abbreviated to—what else?—"CHESSE." The guidelines naturally seized on this connection, contriving for young readers to "travel around to look at Chessie's friends, the CHESSE's." The portion of the text dealing with pollution ("Things that worry Chessie") would have had the serpent exploring some rather technical topics, like "toxic pollutants from factories," "non-point source runoff from fields, homes," "grass disappearance," and "sedimentation." Chessie would reassure readers that the situation was not irreparable and that they could help with cleanup efforts.[27]

Using these guidelines, Harms prepared a series of short prose captions to accompany suggested artwork for each page of the book. Initially, she planned a 20-page book, but the proposed length must have been rejected, because a condensed, 10-page version soon replaced the original. It removed Chessie's individual meetings with bay wildlife, and cut down the section dealing with pollution, but changed little else. Both versions of the book were pretty joyless and had Chessie presiding over a bay on the brink of disaster, with slim hope of revival. An illustration suggestion for page 2, for instance, would have shown the monster "with a somewhat worried expression," with accompanying text telling children, "The Bay is in big trouble and it needs your help." Chessie's discussion of pollutants included a disconcerting line about "fish, crabs and oysters . . . swimming in a pool full of poison, only they can't get out." At the end of the book, Chessie harangued kids about buying low-phosphate laundry detergent and applying less lawn fertilizer. "Tell all your friends about our problems in the Bay. Get them to help, too!" the monster hectored.[28] Even if the science was accurate and the cause was good, it made for pretty miserable reading. This was hardly a plan for the engaging kids' book Kinser had set out to create.

It is no surprise, then, that Harms drafted a second version of the text, this time composed in a style that more closely adhered to the Dr. Seuss model. Designed consciously as a picture book aimed at elementary readers, with a verse and its related illustration on facing pages, this version was far more whimsical than its counterpart. It drew strongly on the "friendly monster" lore to give kids a friend in the bay who was "not very scary" and had "a story to share." Likewise, the monster's fellow bay wildlife were also its "friends" (not "favorites," as in Harms's first version), and although it admitted to occasional annoyance at humans for their polluting ways, it still confessed, "I love them." This Chessie did not condemn, but instead focused on sending a message of hope and voluntarism: time is of the essence, *you* can help, and here is how. With diligence, "week by week," Chessie told kids, *they* could "bring back the rich Chesapeake." Assuredly, *this* was the approach Kinser had been aiming for.[29]

Using both sets of text, Harms assembled two prototype books—mockup #1, drawn from Harms's first, prose version; and mockup #2, from the second, with rhyming couplets—and sent them to office staff to evaluate. There was no contest between the two. Mockup #1 left everyone underwhelmed. Kinser, for instance, noted its "problems with tone" and complained that it was "a little preachy." His strongest criticism, however, was that the project seemed to have drifted away from its original purpose: "We went away from *organisms*," he observed, "go back to CHESSE's." Other staff comments were less pointed but equally lukewarm. It is "more standard (prosaic) and less catchy than the rhymed version," wrote staff biologist Bert Brun. "*Parents* will probably like rhymes better, and they *are important*!" Steve Funderburk, assistant director of the FWS Chesapeake Bay Program, concurred: "I prefer mock[up] #2 although this is well done."[30]

Mockup #2 was ultimately selected for publication, and it is easy to see why. Harms's verse brought a charm and sweetness to the book that took it beyond a mere educational exercise, toward the realm of children's literature. Once the text was married to Folker's artwork, the whole project was elevated into a unified piece that transcended its origins as just another piece of humdrum government literature. Mockup #2 was not just a socially conscious coloring book for kids, it was the culmination of all the cultural trends that had been weaving in and around and through the Chesapeake for the previous 40 years—with Chessie participating simultaneously as mascot, catalyst, cause, and effect. The result was a cultural artifact that was perfectly attuned to the time period of its inception: a book that not only imbued Chessie with a character and personality but also reveled in and relied on the monster's connections to the imagery, history, economy, and even geography of the Chesapeake.

Two spreads are particularly great examples of how the book tied all these threads together. The first introduces Chessie, with the accompanying illustration depicting Chessie rearing up near a shoreline, the Chesapeake Bay Bridge in the background. A bird and puffy clouds complete the idyllic picture. A more immediately identifiable Chesapeake scene could scarcely have been composed at the time. Its simplicity and iconic quality made the artwork a natural choice for the front cover of the finished book as well.[31]

The other notable spread had Chessie teaching readers about the geographical extent of the bay. The illustration here showed the serpent, wearing a mortarboard, snaking around a map of the waterway and its bordering states. It is worth noting that this map underwent some small changes over time. Recent versions have Maryland, Delaware, and Virginia labeled in the same type, along with the bay, on a map that shows the full extent of the estuary and much of the land on either side of it, from the northern boundary of Maryland and points west, to the tip of Cape Charles on the Eastern Shore of Virginia. In the earliest printings, however, Chessie presented a bay that was all about Maryland. The map focused on the portion of the estuary from the Potomac River northward and included labels only for Baltimore, Annapolis, the Potomac, and the bay itself. It even included a causeway intended to represent the Bay Bridge. The omission of Virginia elicited complaints from officials in that state and was swiftly corrected for subsequent print runs, but it remains a telling example of how people—even scientists and officials—unconsciously intertwined the Chesapeake and Maryland.[32]

Once the proof copy was assembled, it went up the chain of command for evaluation and approval. The process was hard going. Although reviewers generally liked the idea of the book and praised the verse and illustrations, the fact that it was ostensibly "about" Chessie led to some consternation that almost derailed publication. "I had upper managers and staff telling me how inappropriate it was for the FWS to recognize a species that does not really exist and the like, plus quibbling with the language of every couplet," Kinser complained in 2005.[33] One of those quibbling upper managers was Donald W. Woodard, assistant regional director of habitat resources for the Boston territory (which had responsibility for the Chesapeake watershed). Woodard recommended seeking out a private publisher: "I have no objection to it being produced and distributed in its present form," he wrote in a memo in April 1986, "but I do have

Chessie poses in front of the Chesapeake Bay Bridge in this page from *Chessie: A Chesapeake Bay Story* (1986).
US FISH AND WILDLIFE SERVICE

some reservations about it being a Service publication. For one reason, I don't believe the Service should even imply to young readers that 'sightings' of the friendly monster are credible. I also think the Service would be subject to ridicule by some critics who may consider this a waste of taxpayer money."[34]

Chessie teaches readers about the bay's geography in another illustration from *Chessie: A Chesapeake Bay Story* (1986).
US FISH AND WILDLIFE SERVICE

Public affairs officer Inez E. Connor was more enthusiastic, commenting that the rhyming verse "conveys an important message in a clear and enjoyable manner." Connor knew the book would draw criticism, "but how better to appeal to and educate this age group," which, she said, a recent study had revealed was "of vital importance for the future conservation of fish and wildlife resources?" Connor advocated for the book to be made widely available and even suggested that Kinser's office approach fast-food chains about using some of its pages for placemats.[35] Seemingly alone among the FWS management, she seemed to grasp what a golden opportunity the coloring book was for the agency and its goals.

As the final touches were applied to the book ahead of publication, the need for a title emerged—and no one seemed to know what to call it. The outside firm that proofed the book provided a long list of suggestions, but everyone, including the project editor who proposed them, Albert C. d'Amato, was lukewarm toward them. All of the titles focused on the bay itself, Chessie (both generically and as a "friendly" monster), the watershed's wildlife, or some permutation of the three. D'Amato favored a short title that "further identifies Chessie," preferably by highlighting the serpent's friendliness. "I'm not sure that profundity or even a marked cleverness is necessary to appeal to the children. Nor do I think that an academic or didactic title is particularly appropriate. The message inside is certainly didactic, but that is, as far as the child is concerned, a by-product, not the initial attraction," he argued.[36]

Although d'Amato reasoned correctly—after all, Chessie *had* been chosen in order to appeal to children—spotlighting the monster in the book's title was still taboo. Eventually, the Annapolis office proposed *Chessie: A Chesapeake Bay Story*, and the title stuck. "It's not all that jazzy," Harms admitted to Connor in late July 1986, "but we'd like to avoid highlighting the fact that a monster may, in fact, live in the Bay. The main idea of the coloring book is that the Bay is in bad shape and people have to clean it up, not that some serpent might live there. Let's try to keep the focus on the Bay."[37]

While his staff was refining drafts of the book, Kinser was exploring publication options. Opposition to the use of Chessie was strong enough within the FWS that the service would not commit to funding the first printing, so Kinser turned to the Army Corps of Engineers. The Baltimore district of the corps and the FWS's Annapolis office often collaborated on projects in the bay region, and Kinser had built a congenial relationship with the leadership there. One day, Kinser visited the corps' office and brought along a draft of the Chessie book. Suitably impressed, they agreed to pay for the first 5,000 copies. Kinser solicited funding from other agencies in the same way, despite official FWS disapproval. "I may have eventually done it illegally because of all the nonsense," a still-irritated Kinser confessed in 2005.[38]

Monstrous Doings

Even as Annapolis office staff were perfecting their own Chessie in the spring of 1986, the "real" monster was making waves again out in the bay. The first week of March, a report came in from the southern end of Kent Island. At first, Robert A. Kenney mistook the creature for his neighbor's black Labrador retriever, before realizing what he was seeing was larger than a dog. He watched the object swim with his binoculars for several minutes, until it disappeared in the distance. The encounter failed to arouse much excitement among the press, but it was a great example of the usual Chessie scenario of a suburban transplant seeing something unfamiliar and calling it a sea serpent. Kenney was, as one local paper reported, a "retired government worker" who lived in one of the waterfront developments in Romancoke.[39]

As it turned out, the Kenney sighting was merely an appetizer for the year's main entrée—an encounter that became one of the more famous in the monster's history. Around 6:00 p.m. on May 25, 1986, the Sunday evening before Memorial Day, Kenneth Boudrie and his friend Jack Bishop were out on Boudrie's dock on the Tred Avon River, near Easton, in Talbot County. What both men spotted heading downriver, toward the bay, they described as brown, serpentine, with "humps" rising successively from the water, about 25 feet long, and 1 foot in diameter. The local paper, the *Star-Democrat*, could hardly contain its glee at a sighting so close to home. "Did ol' Chessie surface on the Tred Avon?" the paper's headline asked. Neither Bishop nor Boudrie was quite sure *what* they had seen, but Bishop felt certain it was "a large animal." "You can quote me on the fact that I've never sighted a UFO," joked Boudrie.[40]

The Bishop-Boudrie encounter inspired much excitement—perhaps more than it really warranted. The sensation around the sighting probably stemmed from the fact that Boudrie and Bishop were local businessmen with little to gain and potentially lots to lose from becoming associated with Chessie. Both men expressed concern that reporting the sighting would make them a laughingstock, but Boudrie, the more retiring of the two, stressed his skepticism in interviews and eschewed the limelight, especially once the initial furor had passed. By contrast, Bishop, a popular Easton dentist, became a local, national, and even *international* celebrity as a result of the sighting and seems to have embraced the role with good humor and enthusiasm. Not only did Bishop appear on the cover of the *Weekly World News*—"He was right up there with the woman who saw the starship," his dental assistant dryly told one reporter—he was also a featured guest, alongside other Chessie notables, on a BBC radio program about sea monsters in December 1986. Bishop even had T-shirts made for himself and Boudrie proudly proclaiming, "I Saw Chessie on the Tred Avon," and later named his boat *Chessie Spotter*. Perhaps it was his destiny to join the Chessie glitterati. The year before, one of his patients, Jan Snead, also saw the monster, although she did not go public about it, and Bishop reported that "many" other patients, "reliable business people," as he put it, had done so as well.[41]

One can imagine Kinser and his staff reading these headlines with satisfaction as they prepared their own version of the monster, conscious of its topicality. Unfortunately, the rest of the year turned out to be less congenial to Chessie. The Bishop-Boudrie sighting marked the end of the line for new reports, and the papers were left to fill the vacuum with serpent-related puff pieces instead. One rather desperate entry came from the *Kent Island Bay Times*, which ran a story in mid-June relating the opinion of island resident Virginia Donaldson, who believed that Chessie was no more than a misidentified mother otter trailed by her young. Donaldson was a relatively recent transplant to one of the island's then-new waterfront developments and, on that basis, might have been expected to endorse the monster's existence. Instead, she rained on everybody's parade. "All of us who have seen something strange in the water are now quite convinced that it's not a monster at all, but a little, furry thing I'd like to hold," Donaldson rather sweetly told the *Bay Times*.[42] Things had come to a pretty pass indeed when even newcomers to Kent Island no longer automatically believed in Chessie!

In truth, though, the cryptozoological part of the Chessie phenomenon was coming to an end by the summer of 1986. In late July, *Richmond News Leader* reporter Rex Springston headed to Easton, Maryland, to follow up on the Bishop-Boudrie story. His resulting article, headlined "Chessie Is Still Splashing Around," put a brave face on the dearth of recent sightings, but in the end Springston had to admit that Chessie was receiving "less fanfare" than in the past. One of his interviewees delivered a blunter diagnosis: "I think people are getting used to it," the St. Michaels, Maryland, resident observed.[43] Ironically, just as *A Chesapeake Bay Story* was heading out to the printer to begin spreading a message of hope and optimism about the bay, the *monster's* story was sinking under the weight of skepticism and ennui.

When the Chessie coloring book finally came out, it was an immediate hit. Initially, it was distributed almost exclusively by the Annapolis office during local presentations about the bay, but as the general public became aware of it, requests for copies began to trickle in. Elementary school teachers in Maryland and Virginia wrote in asking for whole stacks, to be distributed to their classes and used in lessons. Eventually, the office received requests from as far away as Britain and Australia. Other regional offices within the FWS also got in touch, hoping to emulate the book's success in other parts of the country.

As the good press mounted, FWS officials were forced to acknowledge Chessie's usefulness, whether they liked the creature or not—although staffers took care not to ratify its existence. Interviewed by the Annapolis *Capital* in October 1986, Kinser's lieutenant, Steve Funderburk, made it clear that the coloring book was not really about Chessie at all. "When we talk to adults and say Chessie, we're talking about canvasbacks (ducks), striped bass," and other endangered animals in the bay region, he explained. "The legend of the sea monster might interest children enough to teach them about the real creatures that inhabit the Bay." Still bemused by the book's focus on the monster, the paper stressed that its "message is dead serious" and that Funderburk "would like to see Chessie do for the Bay what Smokey Bear did for fire prevention . . . [and] become symbolic of all wildlife in and around the Bay and help clean up the waterway."[44]

In November 1986 Funderburk wrote Dave Folker to thank him for his good work: "Reaction to the 'Chessie' coloring book you illustrated is overwhelming. We are fast running out of our supply. . . . Your imagination has captured the hearts of thousands of youngsters in the Chesapeake Bay area." A similar note went to Harms, commenting on the good publicity the book had brought in. By spring 1987, the FWS was being bombarded by requests for books for community events, and a questionnaire was prepared at about the same time to gauge educators' and students' reactions to the book in order to justify further printings.[45] In all, it is estimated that between 150,000 and 250,000 copies of the coloring book were printed during those early years.

A few months after *A Chesapeake Bay Story* was published, one of Kinser's staff approached him with the idea of building a costume of the monster to wear to community events. Kinser liked the idea: What better way was there to drum up enthusiasm at public events than with a mascot? The first version was rather homemade and primitive, consisting of a papier-mâché head that "weighed a ton and a half" and a felt body costume assembled by some of the staff. Although the original costume was a hit with kids, the physical weight of the headpiece was a major drawback and led Kinser to have a new one professionally made. It cost about $2,500, a princely expense against Kinser's small budget, but an investment he felt was worth making.[46]

US Fish and Wildlife Service Chessie mascot costume.
COURTESY OF CRAIG A. KOPPIE, US FISH AND WILDLIFE SERVICE

The mascot, in its first iteration, debuted at the annual Washington, DC, Boat Show on February 20, 1987. At a specially organized press conference, Frank Dunkle, then director of the FWS, teased attendees with news of a "mystery guest" who would "prove very helpful in saving the Chesapeake Bay and the critters that live in it and around it." After bobbing through the crowd and exhibits in an intentional nod to the real monster's sporadic appearances, the FWS's "new official Bay clean up mascot" eventually joined Dunkle onstage to address the audience. Offstage, Washington, DC, radio personality Jackson Weaver provided Chessie's voice, delivering a short introduction speech and talking about the coloring book.[47] Weaver was a bit of a casting coup to play Chessie, as he was already famous as the original voice of Smokey Bear, a role he played for 45 years. Over the next year or so, he would continue to voice Chessie in a series of public service announcements on his radio station, WMAL, alongside his longtime fellow broadcaster Frank Harden. The mascot reveal colorfully consummated Kinser's ongoing effort to bring increased focus to his office's public engagement efforts. Even the *New York Times* took notice of the FWS's "monstrous doings" in service to the bay. "Chessie will be the centerpiece of a major effort to involve and inform the children of the bay area of the things they can do to help in the cleanup," the paper quoted Dunkle.[48] It was a remarkable reversal of fortune for the monster, after having been shunned by the agency for the previous year.

Demand for personal appearances by Chessie at local events, such as the Sandy Point Seafood Festival and the Bay Bridge Walk, quickly grew. "There were so many festivals going on all the time that that was our way [to educate the public]," remembers staffer Britt Slattery, who joined the Annapolis office in 1988 and frequently donned the Chessie costume as part of her job. "We would take a display and hand out fact sheets and coloring books and wear the costume and walk around handing out the coloring books."[49] Eventually requests for Chessie overwhelmed the office, and staff began loaning out the costume by mail, having two more made and added to the rotation. Requests also came in from other FWS offices, as well as other environmental agencies, asking for advice on creating their own mascots.

In 1988, the Chesapeake Bay Estuary Program was separated from the Annapolis Ecological Services Office to become its own division. Kinser moved over to become the first field supervisor there. Chessie went with him. The CBEP would continue to use the monster to address environmental issues facing the bay as the 1980s gave way to a new decade—even as the actual creature faded into obscurity. By 1992, when Kinser finally left Annapolis, *his* Chessie had virtually replaced the "real" Chessie in the public consciousness.

More significantly, the monster had been refined into the ideal symbol for the Chesapeake and its plight. More thrilling than a skipjack, more attractive than a crab or rockfish, more mysterious than a Chesapeake Bay retriever, at long last Chessie had outswum all of its competitors. What other creature could fill that niche so perfectly? What other icon could seize the imaginations of people all over a region, then convert their interest into concern and maybe even action? Looking back on his time in Annapolis, Kinser felt that local awareness of bay issues had increased significantly, and he traced that improvement, in part, to the use of Chessie in outreach. "I was sitting in a McDonald's one time," Kinser recalled with satisfaction in 2005, "and there was a little girl at the table behind me talking to her father over a Big Mac, and telling what she knew about Chessie, and she started talking about what she knew about the Chesapeake Bay. And I was surprised: you know, here's a nine-, ten-year-old child who has a lot of knowledge. You know it came from someplace."[50]

Chapter 7

Diminishing Returns

EVEN IF NEW SIGHTINGS were now growing scarce, Chessie *the symbol* continued going strong in the two years leading up to the monster's tenth birthday in 1988. Indeed, rebranding the creature as an environmental icon probably extended its life span by distracting the public from asking awkward questions about its credibility. It was enough that the bay now had this singular spokesman, right at the moment when it also had a multitude of problems demanding the public's attention. Chessie no longer needed to make scientific sense in order to make a statement. The monster just had to inspire wonder and curiosity, and hopefully the public's goodwill would do the rest.

One man in Annapolis who was full of goodwill, for Chessie and the bay, was Paul Foer. A yacht captain, entrepreneur, and lifelong resident of the Chesapeake region, Foer was fascinated by reports of the serpent when they became public in the late 1970s, and he eagerly—if skeptically—followed the creature's travels in the papers. When the FWS published its coloring book in 1986, however, what had been a mere curiosity for Foer now became an opportunity. A bit of a showman at heart, he decided to mount an *actual* expedition, of the sort that had been undertaken by famous cryptozoologist Roy Mackal to find the Loch Ness Monster (if on a much smaller scale). Ostensibly, Foer sought to locate Chessie in its secret bay hideaway, but the real purpose of the proceedings was to promote awareness about the state of the bay. It was truly a case of life imitating art, and confirmation that Chessie had become a recognized metaphor for the Chesapeake.

Just Add Water and Mix

Foer's concern for the Chesapeake Bay had deep roots. Growing up in the suburbs of Washington, DC, Foer spent summers with his family on the shores of the bay in Shady Side, Maryland, in the historic home of nineteenth-century waterman and sea captain Salem Avery. The experience left him with one foot planted in the suburban world and one steeped in the life of the Chesapeake—its wildlife, recreational and work uses, history, culture, and folklore. As an adult, he capitalized on this duality. By the age of 19, he was a licensed charter boat captain. In college, he designed his own major, American Maritime and Coastal Studies, based largely on the resources available in the bay region, and studied with George Carey, who served as Maryland State Folklorist in 1974–75. After graduating from college, Foer immersed himself in bay-related work, running a charter service, teaching sailing, and gradually dipping his toes into the growing environmental movement in the region. His concern for the bay's health began at a young age and was directly connected to his lifelong relationship with the waterway. Although he would later be employed with the Environmental Protection Agency's Chesapeake Bay Program in the early

1990s, in the early 1980s his environmental activism mainly took the form of educational efforts undertaken through his yachting business. "I did not perceive the Bay and do not perceive the Bay as that of someone from the urban world who's suddenly come here," Foer reflected in a 2008 interview. "To me, it's my home, it's engrained in me, it's what I've known since childhood, since birth, and I feel very deeply connected with it."[1]

Although he had been aware of Chessie since its first appearances, like most Marylanders, Foer really started to take notice of the creature after the Frew film became a sensation in 1982. He was unimpressed by what he saw in the video, but a possible sea monster in the bay intrigued him. Born in 1959, Foer was a teenager during the height of the 1970s New Age craze. "I was very interested from an early age," he remembers, "in mysteries, ancient civilizations, supposed visits from extraterrestrials—everything from pyramid power to mental telepathy and tarot cards. Always very skeptical, but very interested."[2] Chessie was a similar mystery to unravel, one that seemed, tantalizingly, to offer enough evidence to be worth pursuing at a deeper level than the newspaper headlines. For much of the 1980s, Foer contemplated authoring a book about the monster. In the course of laying the groundwork for the project, he made connections with Chessie A-listers like Bill Burton and Michael Frizzell, as well as notables from the world of cryptozoology, such as Bigfoot and Champ researcher Gary Mangiacopra. The book ultimately never came together, but the research process fed both his interest and his skepticism, encouraging a gently ironic perspective on the Chessie phenomenon.

The Frew video reminded Foer of the inconclusive footage produced by monster hunters searching Loch Ness, and it got him thinking. Nessie had been the object of numerous, high-tech, scientific search efforts and still had not been proved or disproved; yet the expeditions kept happening—irrationally, in Foer's view. So why not capitalize on both the form *and* the absurdity by staging a mock Chessie hunt? "So, I started thinking," Foer recalled in 2008, "Well, we're not going to mount a serious expedition to look for Chessie, so let's mount a tongue-in-cheek one and see what we can do for the good of the Chesapeake."[3] "While he's searching," the Annapolis *Capital* echoed in a promotional article ahead of the event, "Foer hopes to help save the bay—not from the monster, but from monstrous pollution."[4]

By the mid-1980s, Foer was operating a successful charter boat business called Adventures Afloat. Its catchy slogan—"Just add water and mix"—encapsulated what both the enterprise and its owner were all about: namely, connecting people to the bay via boating excursions. Foer took a similar approach to his proposed Chessie expedition, envisioning an event that combined a touristy dinner cruise on the bay, a lesson about conservation, and the sea serpent hunt gimmick. It also drew on some of his business connections: Foer's friend and fellow sailor, captain Tony Fotos, offered his own 60-foot pleasure yacht, the *Lady Anna*, for the trip. The diverse ingredients made for a potent admixture, and Foer was careful to spell out its true purpose. "Chessie is a metaphor for the fragile existence of life in the Chesapeake Bay," he was quoted as saying in the Annapolis *Capital*. "This is just an exciting and fun thing that can get people out on the bay."[5]

The main object of the whole enterprise, at least as Foer envisioned it, was to use Chessie as a hook to raise funds and enlist new members for the Chesapeake Bay Foundation (CBF). Tickets cost $70 per person or $130 per couple for the trip and included a one-year membership in the organization. Despite his good intentions, however, when Foer approached the foundation about taking part, he was rebuffed. "I was told from the highest level on up, 'Well, sure, if you want to raise money for us, we'll take the money. But we don't want you in any way to suggest that this is sponsored by, or in coordination with, the Chesapeake Bay Foundation,'" a still-bitter Foer complained in 2008. Accordingly, Foer took pains to make it clear, particularly to the press, that he did not represent the CBF, nor was the organization sponsoring the expedition. Still, because the foundation was the event's intended beneficiary, the boundary between the two remained murky—a condition perhaps made even more complicated when the *Lady Anna* launched from its berth with a "Search for Chessie—Save the Bay" banner unfurled on its side.

On September 11, 1986, at 7:00 p.m., about 40 passengers, most of them Foer's family and friends, departed the Annapolis City Dock for a three-hour tour of the waters around Kent Island, with special visits to Love Point and Eastern Bay, locations of famous Chessie manifestations.[6] Also aboard were Bert Brun, a US Fish and Wildlife Service biologist who had helped formulate the first Chessie coloring book; Hermann Gucinski, director of the Anne Arundel Community College Environmental Center; and Gilbert Levin, a Rockville, Maryland, engineer who had designed technology used to detect life signs on Mars. Although their presence reinforced the scientific angle of the sea serpent hunt schtick, Brun, Gucinski, and Levin were really invited along to be the evening's guest lecturers.

The evening departure time made for an atmospheric, moonlit trip, and the scientist-presenters aboard brought both scientific legitimacy and a sense of playfulness to the event. Of the three, Levin perhaps entered most vigorously into the spirit of the proceedings. His contribution was a machine he supposedly designed in order to track Chessie's putative radioactive trail—a riff, no doubt, on his reputation for launching a similar device to Mars on the Viking probes in the 1970s.[7] As Foer recalled with amusement, Levin "had this whole fake thing about an iridium detector—you know, with an antenna—because Chessie would have trace amounts of iridium in her body because she comes from outer space. . . . Now, this guy is a multimillionaire owner of a publicly held corporation, and a PhD who's worked on some of the highest-level projects at NASA. He had fun with it."[8]

Not to be outdone, Gucinski supplied hijinks of his own. Armed with scholarly articles on potential sea monster population density in Loch Ness—yes, such things really exist—Gucinski told guests that Bloody Point, the southern tip of Kent Island, was liable to be Chessie's home, because it is the deepest part of the bay. The holes there, he claimed, have plugs at the bottom of them, and if they were pulled—by, for instance, a mysterious sea monster—it would cause the Chesapeake to drain away![9]

Of the three scientists present, Brun's contribution was perhaps the most conservative. Described by the Annapolis *Publick Enterprise* as "a distinguished-looking man despite his 'I Saw Chessie' T-shirt," Brun hewed closer to straightforward environmental messaging than his colleagues, an approach that harked back to the scientific sobriety of the US Fish and Wildlife Service during its consideration of the Chessie coloring book (with which he had been involved). Brun, the *Enterprise* wrote, "was quite serious about the mission's real purpose . . . and he provided a number of interesting and informative pamphlets on particular problems and assets of the delicate ecosystem of the Bay."[10] Brun nevertheless did his part to establish a lighthearted tone for the evening by providing a blow-up pool toy to represent Chessie. Tethered to the roof of the *Lady Anna*, the inflatable looked more like a dragon than the classic Chessie description, but it made for a useful mascot all the same.

Once the expedition reached Bloody Point, the *Lady Anna* deployed a hydrophone and, after rendezvousing with another craft, the *Menehune*, established a "sea monster search pattern." Foer's guests were able to watch sonar monitors and listen to the output from the hydrophone. The crews on both boats claimed to have "picked up a reading that could have been a large creature swimming underneath our vessels" and allegedly recorded audio of the encounter. But everybody knows that sea monster sightings never produce clear evidence, audio or visual—so in keeping with tradition, the tape "somehow was destroyed." As the *Lady Anna* prepared to head back to Annapolis, the expedition had one more surprise in store. John Cece, captain of the *Menehune*, expertly wheeled his sailboat around, drew abreast of the other craft, and threw an envelope aboard with his own donation to the CBF.[11] Although the expedition returned bereft of Chessie—a not unexpected outcome—Foer and his crew were glad to do their part to draw attention to the bay and its problems. Levin figured Chessie felt the same way: "Unfortunately, we didn't get a firm fix on her," he told the papers. "She continues to stay elusive because she knows that as long as she does, she'll continue to attract interest to the Chesapeake Bay."[12]

Even though it failed to locate Chessie's secret hideaway, Foer's outing still succeeded in adding some 30 new members to the CBF's rolls and raised about $350 in donations and membership dues. Writing to Rod Coggin, the CBF's public affairs director, on September 24, 1986, Foer expressed his pleasure at the event's success. "Not only did we bring in new CBF members, raise money and have a good time introducing people to the Bay, we also managed to garner a fair amount of publicity for the Chesapeake Bay Foundation," he explained. Would the organization have any interest in "a similar and perhaps larger event" in 1987? he inquired.[13] Unfortunately, the foundation's leadership remained noncommittal and, Foer believed, ungrateful, and the experience permanently soured him on the CBF. Notably, when the second Chessie hunt took place the next year, Foer found a different beneficiary.

Our Symbol of Caring

Foer began planning for the second Chessie expedition in the summer of 1987.[14] This one would be slicker and more ambitious than its predecessor and was designed to iron out many of that excursion's flaws. The date was set for September 20, from 3:00 to 6:00 p.m. The afternoon itinerary fixed an obvious problem from the year before: namely, that the trip had barely started by the time it was dark—not ideal conditions for serpent watching. Tickets were much more affordable this year, too: $21 per person, out of which a $10 charitable donation would be deducted automatically. One sign of greater attention to the sheen of the final product was the publication of an "official program," the contents of which included acknowledgments, an itinerary, and articles from Foer and other presenters. On sale for $2 apiece, the document unapologetically hawked itself, with a line on its cover declaring that it had "the potential for becoming a great collector's item."[15]

Probably the most notable upgrade for this year's event was the replacement of the comparatively small *Lady Anna* with the 150-foot-long topsail schooner *Clipper City*. In practical terms, the exchange allowed for a much larger number of passengers to take part in the event—125 compared with 40 the previous year—but it also lent the expedition a more classically maritime flavor than the first expedition's motor-driven yacht. A modernized reproduction of a nineteenth-century vessel bearing the same name, in 1987 the *Clipper City* was stationed in Baltimore Harbor, where it was used mainly for local tourism cruises similar to Foer's. In a generous gesture, the ship's owners not only offered it at a sizable discount but also sold tickets for the event. Because the *Clipper City* was berthed in Baltimore, the expedition would depart from there too, touring the Patapsco River outside the harbor and the Baltimore skyline while trying to find Chessie.

A second major change concerned the selection of a recipient for the event's donations. Foer remained gracious toward the CBF when its part in the previous expedition came up, but he bypassed the foundation completely this time around. Instead, proceeds went to the Maryland Watermen's Association (MWA) for its Blue Crab Research and Development Fund. Attendees, the official program enthused, would be helping the association "to study the life cycles and behaviors of the Blue Crab, the most delectable and popular of foods from the Bay. Once thought to be the most resilient of commercial fisheries on the Bay, even the Blue Crab population is showing signs of stress, most likely due to pollution."[16]

The MWA was a felicitous choice on Foer's part, for a few reasons. First, because its constituency was often accused of overharvesting in the Chesapeake, the MWA frequently found itself at odds with purely environmentalist groups like the CBF. If Foer intended to thumb his nose at the CBF, even unconsciously, affiliating with the MWA was a good, if subtle, way to do it. Also, with its history of advocacy favoring work uses of the bay, the MWA's involvement in a Chessie hunt returned the monster to its origins in the conflict over the "correct" use of the water. Maybe the most convincing reason for linking arms with the MWA, however, was that the association actually welcomed the expedition. "I had a good relationship with their president, and with their administrative director, and they worked with me on it, and it was very positive," Foer remembered in 2008.[17]

A pair of familiar faces returned for this year's expedition. Bert Brun and Hermann Gucinski both signed on again as members of the voyage's scientific team. Brun even authored an article that was included in the event program. "Chessie—We Need You" served as both a condensed primer on challenges to the bay's health and an apologia for using Chessie to answer them. "If there really is a Chessie," Brun explained, "she or it would need the same conditions that other creatures need to live in the Bay—clear water, no toxic substances, and sufficient healthy grasses along Bay shores for animal food and habitat and oxygen." The monster was, he concluded, "our symbol of caring" about bay restoration.[18]

New to the crew for 1987 was a familiar face from a *different* team: Michael Frizzell of the Enigma Project. Foer first became aware of Frizzell in the spring of 1987, after he read about the Enigma Project in a newspaper column on Chessie in the Annapolis *Capital*. Coincidentally, the column, called "Chesapeake Corner," was a regular contribution to the paper by Glenn Kinser's staff, and part of his office's ongoing outreach efforts using Chessie. At that time, Foer was toying with the idea of writing a book about the sea serpent, and in April he began corresponding with Frizzell. Although the book project fell into limbo, Foer courted Frizzell hard to join his second expedition. "In my upcoming cruise," he wrote in July, "I would certainly be interested in your input and resources to publicize your efforts as well as to 'lend an air of respectability' as you said in your letter to me."[19] The two eventually met in person in August, and immediately afterward an inspired Frizzell wrote Foer that "the enthusiasm you conveyed . . . was contagious. The more I considered the idea, the more intriguing it became."[20] Accordingly, he agreed to deliver a talk on the monster during the expedition, complete with a collection of "highly relevant visual aids" that included the Enigma Project's copy of the Frew tape. In an era before easy digital reproduction of video, it was a real coup to have access to such an important, and rare, Chessie artifact.

About 100 guests departed aboard the *Clipper City* on Sunday, September 20, 1987. A small flotilla of watermen's boats accompanied the tall ship as it left the harbor. Aboard the *Clipper City*, passengers enjoyed scientific presentations before going to the ship's lounge to watch the Frew film and the videotape of the previous year's expedition. During this time, the workboats began a mock search for Chessie in the area around the *Clipper City*. Foer told attendees, "Rub elbows with our scientific crew, but keep those cameras handy!"[21]

Ironically, however, the part of the Patapsco River where the expedition undertook its "search" was probably the last place a monster, or any other creature, would voluntarily inhabit. Loaded down with floating trash and industrial and sewer effluent, the greater Baltimore Harbor was a poster child for the ills of the Chesapeake in the 1980s. As Voice of America presenter Phil Murray stoically described the scene, "The city skyline—high-rise buildings, churches, and homes—recedes in the distance. The coast is now lined with factories and drydocks. Parents and children line up along the railing and peer into the Bay for anything unusual. What they see, mainly, are empty soda and beer cans, paper plates and cups, and other trash."[22] The irony was not lost on attendees, one of whom flatly told reporters, "If Chessie were around, she wouldn't be in this part. This is the worst part of the bay."[23]

As it turned out, the area was *exactly* where Chessie was going to be found. At the end of the evening, a 24-foot inflatable sea monster and a large inflatable crab were auctioned off to generate more donations for the MWA. The blow-up sea monster was then tied to one of the workboats and towed back to the Inner Harbor, making it clear that this year's expedition had not just raised money and awareness, it had also even found the monster!

Whatever else it accomplished, Foer's second expedition provided an important platform for the MWA to advance its rationale for why bay restoration was necessary. In the past, its work-based conservationist message had been drowned out by the louder and more generic environmental movement in the Chesapeake. Indeed, watermen were often cast as villains at odds with that movement's goals. MWA president Larry Simns alluded to that stereotype when he told a reporter on the *Clipper City* that no watermen he knew had ever seen the monster. "Chessie is probably scared of a waterman because she knows they will try to catch her and sell her," he joked. The real problem was not watermen, Simns told the *Kent Island Bay Times*, but population growth. "People don't realize what they do to the bay," he explained. "And if events like these can raise public awareness then the bay's health can get better."[24]

All in all, the 1987 expedition raised $800 for the blue crab research fund. In October, an appreciative Simns thanked Foer for his "special effort" to help the MWA. "I know the folks aboard the *Clipper City* on September 20th went away with more of an understanding of the problems that are faced in the Bay clean up effort and they had a good time doing it," he wrote.[25] Although Simns expressed interest in working together in the future, a third expedition never took place. The 1987 expedition turned out to be Foer's last. By the end of the decade, Foer was turning his attention toward graduate studies and jobs in the public sphere. There was less time available to devote to projects like the Chessie hunts, especially when the return on that time was comparatively small. Also, Chessie itself was by then a fading sensation. With fewer sightings reported by the end of the decade, the monster was falling out of the public eye.

Many Happy Returns

To the dismay of the Chessie faithful, there were no manifestations to speak of in 1987, or at least none that became public, and 1988 proved a lean year too. By now, after a decade of sensationalized agnosticism, the press seemed to have adopted the attitude that Chessie, if it ever existed at all, was just the Chesapeake's quaint little contribution to the offbeat "science" of cryptozoology. Whatever the validity of that much-maligned field, as the papers seemed to indicate, Chessie was still a poor choice of specimen to hang its hat on. No fewer than three separate newspapers devoted column inches to surveys of cryptozoology in 1988, and these articles' publication bracketed the most underwhelming round of sightings on record. It was an ignominious way to celebrate Chessie's tenth birthday.

The *Richmond News Leader* opened the season in May 1988 with an article by Rex Springston surveying the "field" of cryptozoology. The story naturally led to discussion of the Chesapeake's own contribution, and that, in turn, led to speculation about the lapse in sightings since 1986. Asked for comment, Michael Frizzell perceptively replied, "The media are burned out on this."[26] As if answering a dare, just two days later, a sighting came in from near Easton, Maryland. The *Star-Democrat* reported that on the evening of April 20, Pat and Rosalie Patterson and their young grandson, Jimmy, had witnessed a log-like object approaching the dock of their waterfront home near the mouth of the Choptank River. The three went out to investigate, studying the creature for about 10 minutes. It seemed to react to their presence, moving farther off shore, although the family also felt it was curious about them. Rosalie Patterson told the paper, "We really got the impression that he knew we were watching him." The family claimed that the creature reared its head above the water, and they decided it looked very similar to a dinosaur's head. Unusually, it seemed to snort—not breach but *snort*—several times, a sound that the Pattersons described as "similar to one a person makes when clearing their nose while swimming."[27]

If the snort was atypical for a Chessie description, the way the account was presented in the paper was even more peculiar. Splashed across the front page was a photo of the Pattersons' pier with a bizarre, plesiosaur-esque depiction of the monster seemingly pasted into the scene. The caption read, "An unusual creature fitting the description of Chessie appeared near this spot." Although some highly unusual, and even occasionally a little frightening, artwork had accompanied Chessie reports since the beginning, this was the first time a photograph had been *altered*. And for what purpose? An inset with an arrow pointing to the spot in the photo would have accomplished the same end. Moreover, the Pattersons' description of their monster, especially the snort and the long neck, departed significantly from the typical Chessie sighting. The presentation lent some credence to Frizzell's point. When it came to Chessie, was the press even trying anymore? Or was it trying just a little too hard?

On May 20, 1988, famous Scottish Nessie hunter Adrian Shine came to Williamsburg, Virginia, to help amusement park Busch Gardens celebrate 10 years of its Loch Ness Monster rollercoaster. In another strange Chessie coincidence, the ride had opened to the public on June 6, 1978—about five weeks before the very first monster sighting took place some 40 miles away. The *Richmond Times-Dispatch* sent a correspondent to Busch Gardens to interview Shine, and naturally the Chesapeake's hometown sea monster came up in the conversation. "I find it encouraging that things are seen elsewhere. It doesn't mean, of course that the same thing is being seen," the bemused Scot told the paper. Shine was perhaps being gently diplomatic. His investigations into strange sightings on Loch Ness were vast production numbers involving fleets of sonar-wielding watercraft, the envy of the entire cryptozoological community. Shine himself was one of the glittering stars in that community's firmament, practically sea-monster-hunting royalty. Asking him about Chessie, especially after a two-year sighting dry spell, was the cryptozoological equivalent of asking a major-league baseball player about the prospects for that year's local little league team. Asked if Chessie was "real"—whatever that even meant after 10 years of unconfirmed reports—the genial Shine, whom one suspects was used to these sorts of questions, graciously but agnostically replied that the world's waterways are "very big and there is no way I would dismiss anything."[28]

All the same, the dearth of sightings as the summer of 1988 drew to a close seemed to suggest that the Chessie phenomenon had run its course. With nothing new to report after such a long time, the media turned its attention to other sensations, and the Chesapeake's homegrown sea monster gradually drifted into obscurity. For the next few years, its appearances in the papers would take place exclusively in retrospectives and "news of the weird" types of features. The Annapolis *Capital* got there first with an October 1988 article examining sightings of both Chessie and, perhaps unexpectedly, Bigfoot. "Do They Exist?" the headline wondered. Despite its fame, writer Neff Hudson concluded, Chessie "can be described as the Rodney Dangerfield of sea monsters. It don't get no respect."[29]

Fortunately, the decline in "actual" Chessie sightings had little effect on the monster's ability to play an ongoing role in environmental outreach. By the time of its tenth birthday, the monster's identity as the spokesman of the US Fish and Wildlife Service (FWS) for the Chesapeake was completely assured, virtually to the point that the FWS Chessie existed independently of the "real" monster. An index to the monster's importance was its appearance in the Annapolis office's regular column, "Chesapeake Corner," in the Annapolis *Capital*. On two separate occasions in 1987, April 16 and December 17, Chessie was featured in the column.[30] In both cases, the monster served its usual function as a proxy for more serious matters, but it was clear that Chessie was the cornerstone of the agency's outreach efforts.

In 1988, the Chesapeake Bay Estuary Program (CBEP) enlisted Chessie for another major venture, this time as the anchor for a curriculum designed for elementary educators. *Bay B C's: A Multidisciplinary Approach to Teaching about the Chesapeake Bay* was the brainchild of a then-recent hire at the Annapolis office, Britt Slattery. "I don't know if somebody else put the idea in my head," Slattery remembered in 2005, "but I know that I was trying to come up with some kind of reading component, and we had the coloring book already, and it was popular as a reader. And so it was really just kind of a natural progression."[31] Unlike its predecessor, which was a collaborative effort, this second Chessie book was almost entirely Slattery's creation. An accomplished artist, she was able to mimic the style of Dave Folker's art from the original, drawing new scenes to accompany the text she wrote. The result was *Chessie Returns!*, a sequel to *A Chesapeake Bay Story* that brought the monster and its message up to date. *Chessie Returns!* was pitched at a slightly older reading level than its predecessor, so kids who had first met Chessie earlier in elementary school could encounter the serpent again in later grades, this time engaging with a more sophisticated prose style.

Chessie's lesson had grown up with its readers too. Whereas before the monster had complained about pollution in pretty general terms, now it stressed development—industrial *and* residential—as the main culprit behind the bay's decline. "They tore down more trees and they cleared off the ground / where the animals lived, where their food had been found. / And soon where a babbling brook once had been, / came the buzzing of progress, a deafening din," Chessie narrated over a drawing of a waterfront cityscape. Another page showed a bulldozer digging on a shoreline, next to a sign announcing, "Coming Soon! Best ever Waterfront Homes." People "ripped up the marshes to build close to the shore," Chessie commented. They "kept plowing down trees" because "they wanted more roads and more buildings built here."[32] It was certainly a much more pointed criticism than the monster had previously offered.

Like many sequels, *Chessie Returns!* failed to capture the same lightning in a bottle as its predecessor. Glenn Kinser was Slattery's boss, and he gave her the go-ahead to work on the update. "I don't think it has the simplicity," he observed of the book in 2005.[33] Slattery's reflection was similar, despite her role in its creation. "I think it pales in comparison to the first," she said.[34] They were being a little too hard on themselves. *Chessie Returns!* boasted imaginative verse and quality illustrations. Where it may have taken a wrong turn was in the tone and density of its message and artwork. The first Chessie book had been designed to deliver a light and optimistic lesson in a way that kids would find alluring. Kinser and Jamie Harms had consciously steered away from text and art that might convey hopelessness or, even more important, preachiness. The sequel, by contrast, allowed the monster to get up on its soapbox and point flippers—and, in the process, took away some of the fun that Chessie was supposed to introduce in the first place.

Dead Wake

Chessie's story began winding down after its tenth birthday. With no new eyewitness reports to provide buoyancy, the monster, at least in its biological sense, foundered. The Chessie faithful have never successfully explained what happened to the creature. Did it die? Did it leave the bay? Had it ever really existed in the first place? It remains as enigmatic in its disappearance as it was in its appearance. By the turn of the 1990s, the monster had slipped below the surface of the local zeitgeist: dormant, perhaps, but still out there, waiting. The press kept angling for a return, publishing periodic retrospectives on the monster in the hope of jogging people's interest in their friendly neighborhood sea monster.

One of the most impressive displays in this direction was a multipage, illustrated spread in the September 1989 issue of *Mid-Atlantic Country*. Alongside a couple of moody shots of the Chesapeake—against one, Robert Frew brandishes a camcorder like a gunfighter—the article brought the Chessie phenomenon up to date but had few answers. What made author Tim Sayles's take on the monster different from earlier efforts, however, was an acknowledgment of its ambiguity. In Chessie's heyday, reporters rushed to resolve contradictions in eyewitness accounts, in order to present a coherent description of the creature. Sayles embraced the inconsistencies, although he continued to regard the accounts as "remarkably uniform." Likewise, he almost reveled in the riddle of the Frew film—it could so easily provide the answer to the mystery of Chessie, but that answer lay tantalizingly out of reach. Sayles's article leaves the impression that the ambiguity—the sheer open-endedness of Chessie—was the attraction. "That is the essence of Chessie at the moment," he wrote. "Hundreds of perhapses, a handful of probablys and not a single definitely."[35]

This lack of certainty probably spelled Chessie's doom over at the FWS too. It was an ironic twist: earlier in the decade Chessie's mythic, almost innocent, nature had made the monster an ideal spokesman for the Chesapeake. In spite of its ubiquity there for the previous few years, times, attitudes, and personnel were changing. CBEP outreach programs were becoming more formalized and scientific, and there was less room for frivolities like the bay's friendly sea monster, especially when the bay's status seemed more dire than ever and widespread public engagement with environmental issues appeared to be ebbing. In straitening times, the FWS could less afford innocence.

The monster's last appearance in print for the agency reflected this gloomier outlook. Published around 1990 in partnership with the Chesapeake Bay Foundation and the National Fish and Wildlife Foundation, the FWS's *Bay News 2020* was a 12-page mock newspaper designed with two front pages. On one side, headlines celebrated the bay's return to good health. On the other, they mourned the estuary's destruction. Articles within reinforced each front page's position. One of the pessimistic headlines was "'Chessie' Found! . . . Dead." The monster's lifeless body, the article reported, had been discovered on the shores of Annapolis, "approximately 100 yards from the Chesapeake Bay Restoration Program office," after it had finally succumbed to pollution. The grim accompanying illustration was a far cry from Dave Folker's coloring book art from a few years before.[36]

Chessie's fictional death in *Bay News 2020* anticipated the creature's gradual ousting from the FWS, a process that was charted in real time by the agency's annual progress report documents in the early 1990s. Chessie warranted several references in the CBEP's first report in October 1990, but just a year later, the monster's presence was reduced to a list of FWS educational titles, of which the coloring books were a subset.[37] In a development that probably pleased many in the upper echelon of the FWS leadership, art had returned to imitating life. With no new sightings to fuel the larger Chessie phenomenon, the monster's promotion as an environmental advocate likewise declined. Chessie's days as the common denominator of CBEP outreach had come to an end, just as the monster's manifestations had also done.

Out of all the Chessie initiatives at the FWS, the one that held on to its currency the best was Chessie's personal appearances in costume, a practice that lasted at least into the early 2000s.[38] It is no coincidence that the FWS Chessie with the longest lifespan was the one that literally embodied the creature. Coloring book illustrations certainly encourage the imagination, but there is no replacing the real thing. Even if sightings had stopped and the FWS had moved its messaging on, that costume still connected people—kids especially—to the bay and its mystique. Whatever else Glenn Kinser and his staff had accomplished, they had brought Chessie and Chesapeakiana vividly to life.

Yet even as the FWS and everybody else were predicting Chessie's demise, that bay connection was paving the way for another leg of the monster's journey—into the realm of children's literature. In the early 1990s, the monster starred in two separate children's books: *Chessie: The Sea Monster That Ate Annapolis!* (1990), by singer-author Jeffrey Holland, and Margaret Meacham's *The Secret of Heron Creek* (1991). Although the books targeted different audiences, they both capitalized on the deep connection the monster had with Maryland and the bay.

Holland's work, a picture book similar in style to the FWS publications, positively reveled in Chesapeakiana, featuring a waterman protagonist named Captain Dan, an Annapolis setting, and the surprise revelation that Chessie was a giant Chesapeake Bay retriever! These riffs on local icons were, of course, intentional: in 1984, Holland founded a local folk band called Crab Alley with friends Jane Meneely and Chris Noyes; during its 10-year career, the group toured around Maryland singing, as one journalist put it, "for audiences interested in hearing about the bay, its history, and its plight."[39] *The Sea Monster That Ate Annapolis!* grew out of a song Holland wrote for Crab Alley that, like the FWS books, was designed to entertain children while still educating them about the bay. "You know, you're always looking for a fun thing to play with," Holland recalled in 2005, "and certainly the concept of our very own sea monster is a great theme. And, you know, looking for a way to commemorate our legends and lore . . . that was perfect."[40]

Pitched toward an older age group, Meacham's *The Secret of Heron Creek* naturally delved into deeper themes. In the novel, two Eastern Shore boys endeavor to save Chessie from the clutches of a rich outsider, Mr. Harrigan, who hopes to capture and sell the serpent. Meacham, a Pittsburgher by birth but a longtime resident of Baltimore, eschews much of the typical Chesapeakiana imagery, but nonetheless she draws on the related trope of the idealized Eastern Shore, a region she described in a 2012 interview as "fairly unspoiled."[41] In Meacham's marshy Eden, watermen are virtuous because in tune with nature, water is emblematic of nature's majesty and innocence, and Chessie symbolizes water and the diversity of life in it. We are told, for instance, that the "whinny" the monster emits when William, the book's protagonist, encounters it is instantly recognizable as belonging to a sea creature. "It was a sound,"

Meacham writes, "that made William think of ocean waves crashing, and of ripples lapping against the shore. In that one sound, William heard both the power and the beauty of the sea."[42] When Harrigan harasses and captures Chessie, he is tampering with nature itself and proving himself unworthy of the access to this paradise that his wealth affords him. In this unassuming children's book, was there an implicit criticism of the dominion over nature that affluent suburbanites often claimed for themselves?

It is difficult to know how children received these books, of course, but looking back, it is easy to see them as a culmination of the kind of culture that bay advocates had hoped to generate around the Chesapeake. While the FWS and other agencies might have fretted over the future of the bay's health and declining civic involvement to rescue it, here was evidence of a grassroots conservation ethic emerging in the region directed especially toward kids. Moreover, it was an ethic rooted not just in ecology and science but also in history and culture. Could there be any stronger validation for boosters like Glenn Kinser and Paul Foer, who worked hard to build the connection between Chessie and the environment, or even straight-up monster hunters like Bill Burton and Michael Frizzell, who laid the foundation for that relationship by recognizing the sheer wonder of the bay's majesty? Even the media—if it could stop focusing on the sightings it did *not* have and focus on the phenomenon it *did*—could take some credit for midwifing this serpentine Proteus of the Chesapeake Bay.

Whatever the whereabouts of the "real" Chessie, it is this transcendent, imagined Chessie that has gone down in history and lived on to inhabit the dreams of Marylanders and people living all around the rim of the Chesapeake. Fifteen years after its first appearances, Chessie no longer needed a physical presence to ensure its continued existence. A maritime mobius strip, Chessie simply *was*—and, as far as the general public was concerned, had always been—a fact of life as basic as the bay itself. It was a remarkable reversal of fortune: the sea serpent that had once been an interloper in the Chesapeake was now right at its heart. Skeptics gloating over the monster's demise might well have asked themselves a searching, existential question: If Chessie really did not exist, did the bay?

Chapter 8

Stand by Your Manatee

IN 1995, longtime Chessie hunter Bill Burton penned his own retrospective article for *Chesapeake Bay Magazine*, titled "Desperately Seeking Chessie." The piece began with some soul-searching: After a decade of solid appearances, where was the creature now? Burton proposed a few answers to the question, some tongue-in-cheek, but a larger question loomed that he and the Chessie faithful eventually had to confront: "Did Chessie ever really exist?"[1]

Back in the 1980s, Burton had expressed far more confidence in the monster, but early in the new decade, there was good reason to wonder. In the fall of 1994, the mid-Atlantic had thrilled to the news that a Florida bull manatee had migrated into the Chesapeake Bay. This unexpected behavior surprised biologists, who speculated publicly that the animal, and others, might have been visiting the bay for years without anyone realizing it. Fearing the manatee would die in the cooling autumn waters of the Chesapeake, the US Fish and Wildlife Service (FWS) mounted a rescue operation and returned it to Florida. During that operation, they nicknamed the manatee Chessie, in honor of the putative sea serpent, and the name stuck.

Alas, so did the intimation that the monster might *always* have been a manatee, a proposition Burton rejected. "Associating the original Chessie to a manatee is like comparing the proverbial oranges and apples. Descriptions of the two don't match," Burton grumbled. Although he conceded that the Guthrie sighting in 1980 was probably a manatee, nearly every other eyewitness description he had collected pointed to one conclusion only: "A manatee Chessie is not."[2] Fate, however, was about to prove Burton wrong and, in the process, give Chessie a whole new lease on life.

Chessie Lives!

Over Labor Day weekend 1994, a manatee was spotted swimming in a tributary of the Chester River, near Rock Hall, Maryland—not terribly far from Kent Island, site of so many earlier Chessie appearances. Alerted to the manatee's presence, the FWS asked boaters to keep an eye out for the animal. "The reason that we're interested in the one in the Baltimore area," the agency's manatee recovery coordinator told the Associated Press, "is it's about as far north as we've ever had reported. But with winter coming on, it will not survive." If it continued to appear, then the agency would consider catching and repatriating it before colder weather arrived.[3]

As luck would have it, the manatee did reappear, this time at—where else?—Kent Island. On September 10, boaters reported seeing the manatee floating in a marina on the island. Although one woman mistook the docile animal for a "creature" that "scared her a little," there was no chance that Chessie would be invoked because the owner of the marina recognized it as a manatee, having seen examples before in Florida. The FWS and the Maryland Department of Natural Resources continued to entreat boaters to be careful in the area of the island but otherwise waited to see if the manatee would head back south on its own.[4]

The concern for a single manatee specimen, even if in foreign waters, might seem excessive, but in 1994 it was big news, and not just because the animal was far from home. The West Indian manatee (*Trichechus manatus*) normally lives in the warm waters of the southeastern United States, the Caribbean, and as far south as the coast of Brazil. Docile herbivores that are known for grazing in waterways, manatees are especially vulnerable both to habitat change and to injury from watercraft—trends that by the 1990s had reduced estimated US populations to only about 1,500. They were protected from hunting under the Marine Mammal Protection Act (1972) and listed as endangered under the Endangered Species Act (1973), and by the end of the 1980s, the FWS, the agency tasked with managing manatee populations in the United States, had stepped up its efforts to track the animals. At the state level, in 1975, Florida designated the manatee as its state marine mammal, and in 1981, singer Jimmy Buffett and then-governor Bob Graham established the Save the Manatee Club in order to advocate for the gentle animals. In yet another case of sea serpent serendipity, then, the Chessie furor overlapped with this sharp rise in awareness about a verifiable ecological problem.

Unsurprisingly, it would not take long for these two gentle but mysterious sea creatures to become intertwined. To some extent, they already were: although Burton generally pooh-poohed the notion that Chessie had ever been a manatee, in truth, there is ample evidence of confirmed or suspected manatee activity in and around the Chesapeake Bay, especially its southern reaches, in the late 1980s and early 1990s.[5] No doubt they came earlier too: the Guthrie sighting in 1980—Burton's exception to prove the rule—is a dead giveaway once the possibility of manatees in the bay is admitted. A comparison of Trudy Guthrie's drawing and description of her creature with a photograph of a manatee quickly reveals obvious similarities, and even at the time, the animal was floated as a possible explanation for the sighting.

The growing awareness that manatees were in the bay and its tributaries provided a story that paralleled Chessie the sea serpent's but on a smaller and less flashy scale. Nonetheless, the manatees spotted in the bay followed a similar arc in the press, going from exotic alien interlopers to providing an easy shorthand for the region and its environmental worries. When a manatee apparently wandered into waters near Hopewell, Virginia, in 1987, for instance, Mayor Barbara A. Leadbetter dubbed it "James Mattox," after the two rivers that intersect at the city: the James and Appomattox. If a manatee had swum up to Hopewell, Leadbetter quipped, the two rivers "must be a lot cleaner than some people say they are."[6] The hopeful remark was prescient: conservation authorities would say the very same thing about Chessie the manatee when he arrived in the Chesapeake a few years later.

By the end of September 1994, officials had resolved to capture the rogue manatee, take it for medical evaluation at the National Aquarium in Baltimore, and then return it to its home waters or to Sea World, in Florida. Although the situation was not yet urgent, time was marching on; by later October, falling temperatures would make the estuary too hostile for the manatee to survive. Realizing its staff were not manatee experts, the Maryland Department of Natural Resources leadership called in two experts from Florida: James Valade, an FWS biologist, and Steve Lehr, a senior animal care specialist from Sea World. But how to find the animal? Although signs indicated it had settled into the waters around Kent Island, searching even that relatively small area was a daunting task. Some 60 volunteer divers from the National Aquarium had been searching for the manatee for a week, hoping to cordon off an area that the animal could be forced into and then captured by hand. But when the animal sensed danger, it would submerge into boat channels to escape. As Lehr told reporters, "He knows exactly what he's doing, and unfortunately, he's more suited to his environment than we are."[7]

After about a week of increasingly frantic searching, officials located and netted the manatee in Queenstown Harbor on October 1. Valade believed the animal would be returned to Florida within a few days, and—because motorboat injury scars on a manatee's back are as characteristic as a fingerprint—probably would be returned to the specific area he came from. As it turned out, the wandering manatee was released a week later, on October 9, in the Merritt Island National Wildlife Refuge near Kennedy Space Center. The cost of transportation was financed with federal funds set aside for this purpose, and by the Save the Manatee Club.[8]

News of the manatee's rescue was met with elation. Like Chessie the sea serpent, from its first appearance, it had seized the public imagination. When rescuers finally netted and loaded the animal into a truck bound for the National Aquarium in Baltimore, onlookers applauded. "He captured our hearts from the outset. Now he's headed back home," one newspaper enthused. The manatee had such currency that a few days later realtors actually invoked it in an advertisement for a Kent Island housing development. "Manatee Enthusiasts: Open Sunday, 1–4," the ad trumpeted. "The Manatee found a safe harbor in Queenstown. You can too!"[9]

A Big, Aquatic Teddy Bear

An unsung hero of the 1994 manatee operation was Kathi Bangert. Bangert had cut her teeth at the FWS working for

Glenn Kinser at the Annapolis office, where she had been steeped in all things Chessie. By 1994, she had risen to become the assistant supervisor in charge of outreach and education there. "I was essentially the public affairs officer," Bangert remembered in 2005. Once sightings of the manatee were confirmed, Valade made contact with Bangert in order to coordinate FWS efforts between Maryland and Florida. Because of substantial public interest in the manatee story, it was important to keep the media informed; it was also necessary to keep the public from interfering with the search.

Bangert had the honor of christening the manatee Chessie. In the aftermath of its return to Florida, Valade asked her to provide the animal with a name for documentation purposes. "He gave me the privilege of naming it," she remembers, "and I said, 'Well, Chessie, of course.'" The selection had less to do with the monstrous sightings of the 1980s than it did with the prevailing optimism of the time concerning the Chesapeake Bay's environmental recovery. "I chose the name just so that we would have sort of a real-life mascot for Chesapeake Bay," Bangert explains. "And of course we talked it up like it came to the Chesapeake because . . . the SAV [submerged aquatic vegetation] was recovering, and it, you know, had a lot to eat, and really enjoyed the waters of the bay." At the time, hopes were running high that the Chesapeake was rebounding and that the restoration efforts of the previous couple of decades had not been in vain. Chessie the manatee symbolized the consummation of the work that Chessie the sea serpent had begun. And if a manatee seemed an odd choice to represent the health of a mid-Atlantic estuary, surely it was no more unusual than a mythical sea monster? "It was an opportunity that presented itself," Bangert concluded pragmatically, "and people just seemed to be so interested, and, you know, really drawn to this manatee that I guess I just took advantage of the opportunity."[10]

Bangert's choice of name was a stroke of genius, because it instantly plugged Chessie the manatee into the zeitgeist already generated by the sea serpent. Not everyone, however, was inclined to regard this new Chessie as a symbol for environmental recovery. The day after the manatee was released back into the wild in Florida, the Annapolis *Capital* ruefully chided Marylanders for their explosion of interest in the manatee but a lack of corresponding interest in the bay. "We are a droll species," the paper grumbled.

> Marylanders are entrusted with a bay that is one of Mother Nature's crown jewels, but we can't agree on how to preserve it and its wildlife—and our attention frequently wanders away from the problem. But let someone else's wildlife take a wrong turn into our bay, and suddenly it's a front-page story, a subject of radio ballads, and the focus of an all-out effort by assorted federal, state[,] and private rescuers, plus volunteers. Go figure.[11]

Though the *Capital* might scoff, Chessie the manatee had earned a small but permanent berth in the ecology of the Chesapeake and in the imaginations of locals. The papers continued to carry stories about the manatee's movements down in Florida, including the smutty news that he had lost his tracking transmitter during some "frisky behavior with a lady manatee"! And Chessie was quickly invoked to help justify environmental projects, such as a land preservation effort under debate in Anne Arundel County in November 1994. At Christmas, Chessie the manatee was featured as an entry in the annual Lights Parade at Annapolis Harbor, a sure sign of the manatee's integration into local lore.[12] As novelist John Barth, no stranger to the Chessie legend himself, remarked in a *Sun* column on Christmas Eve, the manatee was "evidence that truth can be, if not stranger than fiction, more ecologically encouraging." "Next year, giant sturgeons! Maybe even oysters!" he joked.[13]

After Chessie was returned to Florida, there was some debate in the press about the possibility that the manatee might make a return visit to the area, but officials remained agnostic. Valade confessed to reporters, "He's certainly familiar with the route. We just don't know if he'll try it again."[14] As it turned out, the wait to find out was short. On June 15, 1995, the now-retagged manatee embarked on a 19-day, 550-mile trip northward, appearing near Norfolk, Virginia, on July 4. Biologists tracking his movements were dumbfounded. "Manatees seem to have a need to explore," a bemused Valade explained in July. "They wind up getting into funny places. But this guy here has pushed it to the outer limits."[15]

For about two weeks in July, Chessie meandered around the southern reaches of the bay, from the mouth of the Potomac River to Hog Island, one of the barrier islands on the Atlantic side of the Eastern Shore of Virginia. During Chessie's sojourn at Hog Island, Jim Reid, of the National Biological Service's Sirenia Project, succeeded in attaching a new tracking device to Chessie's tail. The existing transmitter, a replacement for an even earlier one Chessie had lost during his "frisky behavior" several months before, suffered from a couple of deficiencies. To begin with, it was vulnerable to salt water, a fatal flaw if Chessie intended to spend any amount of time in the Atlantic. It also lacked an antenna, such that the manatee could only be tracked when he was swimming extremely close to the surface—a limitation that often made following Chessie's trail a crapshoot. Reid's new transmitter was equipped with an antenna that could send a signal with the manatee at a depth of six feet. The upgrade was well timed, since the manatee now seemed intent on swimming farther up the coastline. By July 20, he had reached Assateague Island, just south of Ocean City, Maryland, and there was no hint he had any plans to stop.[16]

Chessie's return trip to the Chesapeake in 1995 cemented the animal in the public's affections and created a sensation not unlike the one the monster had generated in the previous decade. This time around, it was even easier to love Chessie. This was no mysterious, unidentifiable, serpentine creature, it was a cuddly and lovable manatee that *everyone* could believe in. As Bangert put it, "He's like a big aquatic teddy bear. He's so ugly he's cute."[17]

Like the monster before him, Chessie the manatee was also quickly put to use as a symbol. Editorial pages overflowed with manatee-themed columns in 1995. In the Easton *Star-Democrat*, for instance, Robert Perkins used Chessie to parody immigration issues, warning readers to "beware the manatee precedent." Let one manatee into the Chesapeake Bay, Perkins jested, and the next thing you know, they would take over the neighborhood! "Are our schools of rockfish to be jammed with illiterate aquatic vegetarians that don't even speak our language, and whose parents pay no taxes?" he wrote. "Are our few remaining oysters to be pushed out of their comfortable beds by a horde of thrusting aquatic bullies a hundred times bigger than they? Are our carefree outboarders to have their propeller-blades sabotaged by flabby animals deliberately lying in wait for them in the public waters? If the manatees can dodge taxes, why can't they dodge outboards?"[18] It was not especially subtle material, to be sure, but it was a good example of the manatee's flexibility as a symbol, even for issues outside of environmentalism. In a case of history repeating itself, the identification of Chessie as an invasive species—Perkins explicitly referenced zebra mussels, an invasive species posing a serious threat to Chesapeake oyster production—also harked back to similar accusations directed at his sea serpent predecessor. Ironically, back in 1982, the *Star-Democrat* had lobbied its readership to regard the monster's presence as an "advantage." Times had certainly changed in the intervening years.

Over at the *Kent County News*, in Chestertown, Maryland, one of the *Star-Democrat*'s sister publications, the editors took a more sober view of the animal. After first lampooning and apologizing for the media's over-the-top treatment of the manatee saga, the paper nonetheless conceded that Chessie's survival was newsworthy, at the very least because it was so miraculous. And the greatest miracle—here the paper adopted a more serious tone—was the recovery of the health of the bay that had made the estuary desirable to the manatee in the first place. "The serious side of this otherwise lighthearted summertime yarn," the paper wrote,

> is that a few years ago it likely would not have been possible for a manatee to have happily existed in the Bay, getting fat off its lush grasses. The grasses have made an astounding recovery over the past 20 years, thanks almost entirely to the energy and money that have been poured into the Bay recovery program. Yet that same program is now in danger of serious financial cuts, thanks to Congress. And that takes Chessie's story out of the category of summer froth and puts it into the category of real news.[19]

A letter to the editor in the Wilmington, Delaware, *News Journal* echoed these sentiments. "A manatee recently visited our shores," New Jersey resident Michael Humphreys wrote. "Does anyone care that only 1,800 of its kind are left in the world? Would it matter if there were no manatees at all left? These things and much more are likely if Congress has its way unhindered."[20]

Both of these commentaries were responding to the very real possibility that Congress would slash federal spending across the board, as a result of the conservative "Contract with America" platform that had won Republicans both houses in November 1994. The proposed legislation had targeted nearly every sector of the federal bureaucracy for budgetary cuts. Conservative animus toward the Endangered Species Act and its enforcers was especially strong, so environmental services like the FWS and its allied agencies were high on the list for the chopping block.[21]

In Annapolis, the *Capital* met this second round of manatee gazing with characteristic drollness. "We've always said that once a visitor gets a full taste of the beauties of the Chesapeake Bay, we've got him hooked," the paper observed. "He'll be back. But we didn't realize this rule applied to manatees." The *Capital* editors were not exactly happy about their discovery: "We were starting to think that he'd concluded that a free stay in the National Aquarium—the Ritz-Carlton of the flippered set—and free air transportation home are part of the standard package for tourists in Maryland. Not that we have anything against manatees, but we don't want to keep funding Chessie's summer vacations."[22]

Such sentiments were not just spoken in jest: the price of manatee mania—and not just to taxpayers—was a quiet but persistent concern that reared its head every time Chessie did in 1994–95. In a letter to the *Baltimore Sun* in October 1994, for instance, Anthony G. Girandola Jr. complained about the extravagance of such wildlife rescue efforts. "Don't get me wrong, I like the manatee just as much as anyone else," Girandola wrote. "But come on, how many millions of dollars must we spend on these rescues before we say enough is enough?"[23] The *Cecil Whig*, another sister publication of the *Star-Democrat*, located in Elkton, Maryland, was even less forgiving. After a five-year-old boy was murdered by two older boys in Chicago, the paper wanted to know why the epidemic of crime and violence among children was not being addressed. "Where is the outcry? We did not hear it here. We heard more concern about the safety of a manatee in the Chesapeake Bay," the paper's editorial spat.[24]

Where No Manatee Has Gone Before

Concerned citizens in the mid-Atlantic need not have worried, as the manatee largely bypassed the Chesapeake on his second trip. After the stop at Assateague Island, he followed the coastline northward, past the Maryland and Delaware beaches. By the end of July, he was in Delaware Bay, where he lingered for several days, to the increasing bewilderment of marine mammal experts.[25] On August 1, the papers reported that Chessie had reached Atlantic City, New Jersey. "The Energizer Bunny doesn't have anything on Chessie the manatee, who keeps going, and going, and going," Leslie Crook wrote in the *Star-Democrat*. Biologists were less sanguine: by now they had begun monitoring water temperatures ahead of Chessie, as far north as New York.[26] Now that they knew such a long-distance migration was possible for manatees, they wanted to see just how far Chessie would travel, but they kept a watchful eye over him too, in case he ran into trouble. They assured the public that if the water grew too cold—below about 68 degrees—they would rescue him. "To say this guy's freaky or bizarre or something's wrong with him, that's not the case," Reid told the Associated Press. "He's a healthy adult who's knowledgeable about his environment. We know he made the Florida-to-Chesapeake Bay trip last year. That suggests this is part of his movement pattern."[27]

By the end of the first week in August, Chessie was closing in on the border between New Jersey and New York. Scientists were stunned; no manatee had ever been recorded traveling so far north. They began to worry more and more that members of the public might harm the animal while trying to get a look at him. "The more you all pinpoint it," one official chided the press about its excessive coverage of Chessie's odyssey, "the more it's likely to be hit by boats. It could be a big problem. All it takes is one inadvertent turn by a boat captain to run right over it and kill it." For Chessie to make it all the way up the eastern seaboard only to be killed wantonly in New Jersey would be "awful embarrassing," he warned.[28] It was a sentiment repeated on numerous occasions as the manatee moved deeper into territory where his kind were unheard of.

On August 3, Chessie swam past Coney Island, then the Statue of Liberty and Ellis Island. On Tuesday, August 8, the media reported that he had made it to New York City's East River. Biologists probably sighed in relief when he quickly turned around and left. Manatee food was scant there, and there was plenty of pollution and floating debris that could have harmed him.[29] During his stay in New York, the media had a lot of fun with Chessie. Since experts had posited that Chessie's migration might be part of his mating process, the papers imagined a love story taking place out in the waters off the Big Apple. "After all," columnist Michael Daly wrote in the New York *Daily News*, "Manhattan teems with one-time nebbishes who arrive here still feeling the sting of romantic failure."[30] Officials played up the new, humorous twist in Chessie's story. When asked if the manatee was looking for a mate in the north because he had been "scorned in love" in Florida, Linda Taylor of the FWS preserved his dignity: "I think that's rather personal for us to ask him at this point," she replied.[31] Chessie's visit to New York City was not all fun and games, though. A *Daily News* political cartoon depicting a graffitied manatee waving farewell to a trash-ridden New York Harbor showed that Chessie still had symbolic power—and a sting in his tail.[32]

To the continuing surprise of biologists, on August 12, Chessie was pinpointed near Bridgeport, Connecticut; a trip a little way up the Quinnipiac River, and a short stop in the Stewart B. McKinney National Wildlife Refuge, just up the coast, followed.[33] Columnist Denis Horgan of the *Hartford Courant* applauded the stolid Chessie's arrival in Connecticut. The manatee, he wrote,

> cuts a fine figure, demonstrating all those qualities we treasure in our national character and would have the kiddies learn. . . . He is hardy and adventurous, mostly asking to be left alone snacking. . . . Independent, curious, innovative and no burden to anyone, Chessie probably better represents our goals than even do the wise scientists and shipping magnates who would air mail him back to Florida—a goofy state if there ever was one.[34]

Alas, Horgan's column was printed the very same day that the *Courant* reported Chessie had moved on once again. The manatee, the paper complained, was "a typical Connecticut tourist. The state was not his vacation destination, just a nice spot to spend a few days poking around." Despite a short foray into the Connecticut River, Chessie kept swimming northward, across the state line into Rhode Island. On August 16, surfers spotted him near Watch Hill, and by August 18, he had reached Point Judith, at the mouth of Narragansett Bay.[35]

At that point, the manatee seemed to turn around, returning to Long Island Sound, along the coast of Connecticut. Observers immediately assumed Chessie's long journey had come to an end, and he was now taking the first steps homeward.[36] Then: the shock news that Chessie had lost his transmitter in New Haven Harbor. "Chessie the manatee is off the air—and on his own," the *Hartford Courant* portentously reported. "Maybe Chessie just wanted a little privacy," the *Richmond Times-Dispatch* joked.[37] Still, experts remained confident about the manatee's ability to get home safely. "Sure, it's disappointing," Reid confessed to an Associated Press reporter after traveling to Connecticut to retrieve the transmitter. "Yet overall, I'm real [*sic*] happy that we were able to successfully track his movements north through 11 states."[38]

But where was Chessie? For nearly three weeks, the manatee fell off the radar. Experts asked the public to report any sightings, but none were forthcoming, and the trail went cold. Finally, on September 8, he reappeared near Long Branch, New Jersey; the next day he was seen about 18 miles south of there, at Point Pleasant. "He seems to be doing just fine," Linda Taylor told the press. "We hope he continues to move south as the waters continue to become cooler."[39] Chessie's rediscovery ironically torpedoed the excitement surrounding his incredible journey. Over the next several weeks, articles covering the last leg of his trek slipped off the front pages, before tailing off completely. The manatee's return trip past the Chesapeake, in mid-September, received scarcely any comment in the papers, except for news of a brief stop near Norfolk on September 21 and a brush with drowning in the Great Bridge Lock in the Intracoastal Waterway a few days later.[40]

Without the transmitter to track his movements, no one could be sure when or if Chessie actually made it back to his native waters in Florida. A positive identification was finally made on November 16, when Florida Department of Environmental Protection staffer Bill Brooks reported the manatee swimming in the warm water outflow at the Jacksonville Electric Authority Southside Generating Station.[41] A triumphant press rejoiced at the news. New York *Daily News* columnist Michael Daly called Chessie "Man(atee) of the Year." Compared with the plethora of disappointing public figures that had graced New York City in 1995, Daly wrote, the gentle and dependable Chessie was one that the city could "rightly admire."[42] The *Hartford Courant* enthused, "If you're going to send Chessie . . . a postcard, you can use his Florida address."[43]

In the end, Chessie's big splash in 1995 caused lasting ripples that would have made his sea monster namesake proud. "Chessie has boldly gone where no manatee has gone before," Save the Manatee Club spokeswoman Nancy Sadusky proclaimed in November. "His journey has not only set a new manatee migration record, but he has also garnered support and increased public awareness for the plight of his fellow endangered manatees."[44] Symbolic of this increased awareness, although probably a coincidence, was the US Postal Service's inclusion of the Florida manatee in a collection of stamps recognizing endangered species that year.

Ultimately, Chessie may have helped save the FWS itself. Fortuitously, Chessie's adventure up the eastern seaboard took place at exactly the right moment to help make a very public case for funding the often-hidden work of conservation. When it was finally completed late in the fiscal year, the 1996 federal budget slashed fewer funds than anticipated, but the FWS and its ilk still took a considerable hit. Notably, the National Biological Service, the FWS's sister agency that oversaw the Sirenia Project on which Chessie tracker Reid and others worked, was abolished altogether. The project was subsequently absorbed into the US Geological Survey.[45] Was Chessie responsible for rescuing the Sirenia Project and other conservation programs from being defunded into oblivion? Probably not entirely—but all the good press he generated certainly did no harm.

The Rock Star of Manatees

It is fair to say that by the late 1990s, Chessie the manatee had mostly eclipsed Chessie the sea serpent. Indeed, despite Bill Burton's protestations, the former provided a satisfying and plausible explanation for the latter that quickly became widely accepted. If any doubts persisted on this point, a third excursion by the famous traveling manatee would dispel them come the summer of 1996.

In the meantime, Sirenia Project biologists busied themselves with the work of cataloging and bolstering the manatee population in Florida. On February 21, 1996, Chessie was fitted with a new transmitter as part of this work. Around the same time, the Save the Manatee Club donated $24,000 to support the project as budget cutbacks loomed. The club advertised Chessie as a candidate in its "Adopt-a-Manatee" fundraising program the same month. For a mere $20, Chessie's "adoptive parents" received a picture and biography of the manatee of the hour, an adoption certificate, and membership in the club. Chessie's retagging coincided with another piece of good news: estimates of the Florida manatee population went up by about 800, making a total of approximately 2,600 examples of the species.[46] Even if the Sirenia Project were endangered, manatees themselves seemed to be on the rebound.

Reid and other experts predicted that Chessie would venture north again in the summer of 1996, and, true to form, so he did.[47] On June 13, the manatee was located off the coast of Georgia, and by July 4 he had made it as far as the border between the Carolinas. The papers, if not biologists, seemed certain he would make for the Chesapeake Bay.[48] Maybe he would have. Maybe he *did*. Unfortunately, just about the time Chessie was heading into North Carolinian waters, Hurricane Bertha was too. On July 10, Chessie was reported at Morehead City. Two days later, Bertha made landfall as a Category 2 storm about 80 miles down the coast.[49] In the aftermath of the storm, Chessie went MIA; his transmitter was found at Beaufort, North Carolina, on July 17. Still, Reid and other officials expressed confidence that Chessie was fine and probably still making for points north.[50] Whether that would include the Chesapeake was a matter of conjecture.

Although there were reports of additional sightings in Maryland and Virginia in July and August, no positive identification resulted. The sporadic, unconfirmed sightings, combined with the newfound knowledge of Chessie's migrations, put the press in mind of his sea serpent namesake. News articles increasingly, if tentatively, concluded, as the Annapolis *Capital* did, that such travels "might be behind repeated sightings of a 'sea monster,' also nicknamed Chessie," in previous decades.[51]

Experts also speculated that *multiple* manatees, not just Chessie alone, were migrating to the Chesapeake Bay. David Schofield, the National Aquarium biologist who had helped with Chessie's repatriation in 1994, suspected there were at least two other animals responsible for local sightings.[52]

The ubiquity and celebrity of the manatee gradually became a source of frustration for officials. "Every swirl in the Chesapeake Bay, if people don't know what it is, becomes a Chessie sighting," Reid complained in August, echoing the statements of officials dealing with sea monster reports in the 1970s and 1980s. "Isn't there other news around Washington, DC?"[53] Even the papers voiced fatigue as a confirmed Chessie sighting remained elusive. "Are we the only ones starting to find this insatiable desire to know the whereabouts of a 1,200-pound manatee ridiculous?" the Annapolis *Capital* carped in mid-August. The *Baltimore Sun* expressed its exasperation in a lighter way: "Chessie the Manatee has been spotted in the Patapsco, Prettyboy [reservoir] and Loch Ness," columnist Dan Berger quipped.[54]

Chessie probably continued swimming north every summer afterward, but his lack of transmitter kept his movements a mystery. Five years elapsed before he was positively identified again, at Great Bridge, Virginia, in early September 2001. "We were kind of concerned," a relieved Schofield told the papers. "Manatees are often victims of boat strikes. . . . We assumed the worst—that he might have been killed. We're all very excited that he's back."[55] After 2001 Chessie's trail went completely cold. Subsequent unconfirmed manatee sightings were made around the region and along the Atlantic coast, each time raising hopes for Chessie's return, but it would take another decade for the intrepid manatee to make a definite reappearance.

On July 16, 2011, the *Baltimore Sun* announced in a front-page story that Chessie was back! He had appeared July 12 in Flag Harbor Yacht Haven, a small harbor on the Chesapeake Bay in St. Leonard, Maryland. From the scars on the manatee's back, biologists were able to identify him without a doubt. Locals were delighted at Chessie's return. "There's a special feeling in my heart that the animal has made his way back to the Chesapeake Bay," Gene Taylor, a staffer at the National Aquarium who had been involved with Chessie's rescue in 1994, told reporters.[56] The *Sun* was more succinct: "Welcome back, old friend," the paper enthused in an editorial.[57] Sadly, the reunion was brief and final. Chessie lingered for a few hours, probably grazing on Chesapeake-flavored aquatic grasses, before departing for submerged pastures new. It would be the last time that he would be seen for a long time.

Back in 2001, Reid called Chessie the "rock star of manatees."[58] It was an apt description. No other example of his species has done as much to raise public awareness of manatees as Chessie; few animals *at all* have ever done as much for their own kind as Chessie. As an advocate for environmentalism, and for the promise of conservation programs, he has proved to be a symbol even more durable than his namesake. He has also become a convenient, retroactive explanation for the sea serpent sightings of the 1980s, an elixir almost tailor-made to cure the public's monster fever once and for all. If there is any doubt on this point, consider the FWS's stance on the monster in the last 20 years or so: "Chessie is not a sea monster," the agency told this author unequivocally back in 2005. "Chessie is a Florida manatee that migrated unusually far north, even as far as New England, for several years."[59]

Chapter 9

Familiar Friend

CHESSIE'S SIRENIAN RENAISSANCE IN THE 1990S, while charming, nonetheless allowed authorities to put a fig leaf over the sensationalism of the 1980s and effect a nearly complete reclamation of the creature from, as they saw it, the realm of fantasy. No longer a "monster," Chessie was knowable, quantifiable, and—at long last—familiar (if still a little exotic). Under normal circumstances, the story would probably end there, but as we know by now, Chessie is nothing if not persistent.

Authorities may have apprehended the creature's *physical* form by mid-decade, but its elusive *soul* was still free to swim the waterways of the Chesapeake. As the 1990s came to a close, Chessie had been installed in the public mind as a benevolent, full-fledged component of Chesapeakiana, a piece of bay life as natural, peculiar, and beloved as the crab, skipjack, or retriever. By the turn of the millennium, Chessie the sea monster had proliferated all across the region and around the ring of the bay. Once freed from the narrow confines of scientific plausibility, the monster proved immensely malleable, capable of signaling that essential connection to the bay and its heritage that was increasingly valued even as *actual* connections were pushed aside by consumer culture and development. Localness and watery identity could be achieved as easily as adding the name "Chessie" to a product. Chessie was just too useful and durable an icon to let go of.

And in the ensuing decades, Chessie has gone on to become the region's familiar spirit. More than a mere symbol, the creature is a guide to—and maybe even guardian of—what the Chesapeake has become by the twenty-first century. Bay residents have taken the creature to heart, ensuring it swims on, despite its improbable history. These modern-day manifestations of the monster are diverse and often conflicting, but somehow Chessie continues to fill a niche, saying something essential about the Chesapeake that only a mythical sea serpent can convey. It is an ironic destiny for a phenomenon that began life on the geographic periphery and in the doldrums of journalists' silly season.

Ride the Dragon

The spirit of Chessie proved alive and well at the end of the 1990s, when Maryland fielded an entry, dubbed *Chessie Racing*, in the 1997–98 Whitbread Round the World sailing regatta. Held every three or four years since 1972, the Whitbread was based in Britain, with checkpoints in countries spanning the globe. During the previous race, in 1993–94, local boosters had tried without success to establish Annapolis and Baltimore as the American checkpoint. Four years later, the Chesapeake got its opportunity when George Collins, an amateur sailor and head of T. Rowe Price Associates, a prominent Baltimore mutual fund corporation, was brought in to help drum up support for the race from the business community. After touring the region and meeting with state officials, the race committee agreed to make the Chesapeake a leg of the race—if the region could field its own entry. Collins readily committed, and, as he and press officer Kathy Alexander later wrote, "Chessie was brought to life, by a man of Scottish ancestry, in the form of a Whitbread 60 (W60), a sleek, high-tech racing yacht."[1]

Initially, this enterprise was intended to rely solely on corporate sponsorships for funding, but Collins was able to attract the interest of the nonprofit Living Classrooms Foundation, of which he was a board member. The foundation had been using sailing as a basis for educating at-risk students, and Collins proposed a partnership through which he would provide funds for an educational program that would teach basic subjects like math and history through the course of the race. Because it was a nonprofit venture, the resultant Whitbread Education Project enabled donations from individuals and "gave the crew an additional reason to win—for the students."[2]

Children's involvement in *Chessie Racing* was extensive, even though none were actually aboard the boat. A curriculum was developed that taught typical areas of instruction in conjunction with narrower areas germane to sailing, such as sailboat design. Students "participated" in the race vicariously: they "prepared and tasted freeze-dried meals, learned the different sails and their purposes, and plotted courses from one port to the next." To show off what they learned in class, students competed in the "*Chessie* Chase," a "virtual Whitbread Race" sponsored online by the foundation. The curriculum was eventually picked up for use at over 500 schools, and in preparation for teaching it, instructors took part in a workshop showing them "how to turn their classroom into a boat and participate in the *Chessie* Chase." Part of the workshop included "sailing on the Chesapeake Bay aboard the *Lady Maryland*, a traditional schooner."[3]

But why *Chessie Racing*? Kid appeal and the lure of the Chesapeake united once again to give life to the bay's monster. It was George Collins's wife, Maureen, who named the boat. "If she is representing the Chesapeake," she argued, "why not name her *Chessie* for the Chesapeake's legendary sea monster?" Because of the educational dimension of its partnership, the syndicate wanted "a playful, friendly logo" that would signify accessibility. Maureen Collins suggested "a friendly sea monster," and a staffer's wife "whipped up a rough sketch of just such a character, wrapping it around the globe and putting Revo sunglasses on it."[4]

Alas, *Chessie Racing*'s actual performance in the regatta was modest. The sailboat lagged behind its competitors for most of the race, and when the entries arrived at the Chesapeake checkpoint, Baltimore's Inner Harbor, on April 22, 1998, *Chessie* brought up the rear. While the crew might have been disappointed at this placement, the public's excitement was not dented by it at all. Thousands of spectators greeted the boat's arrival at the Inner Harbor, with banners hung from buildings crying, "Go *Chessie*" and "We love you, *Chessie*." "Baltimore gave us a great reception," recalled one crewmember.[5] During the eight days allotted for the stopover in the city, over half a million visitors—many of whom were schoolchildren who had followed the race online—visited the Inner Harbor's race exhibit. Even though the boats ordinarily berthed in order of finish, race officials yielded to *Chessie*'s popularity and reversed the sequence so the sailboat could be closer to the crowds. As George Collins and Alexander later wrote, "Her fans happily clicked away through rolls of film."[6] It was as though a *real* sea monster had appeared out of the bay!

In an acknowledgment of Maryland's rich sailing history and culture, *Chessie* organizers arranged for the Chesapeake checkpoint to include Annapolis as a stopover as well as Baltimore.

Accordingly, *Chessie Racing* relocated to that city's harbor for a second festival held on May 1. En route, the sailboat was cheered on by crews of watercraft large and small, and even from the Naval Academy. "Annapolitans," wrote Collins and Alexander, "demonstrated their love of *Chessie* and sailing with much gusto." On May 3, 1998, state dignitaries, including Governor Parris Glendening, saw the sailboat off on its penultimate leg, to La Rochelle, France. Tens of thousands of onlookers watched from nearby beaches and the Bay Bridge as she got underway. The route out into the bay was lined with thousands of watercraft, most of which were flying the *Chessie Racing* banner. "As far as the eye could see down the bay, a big wide space was bordered by fluttering green flags," one of the *Chessie* staff later recalled with affection.[7]

Although the sailboat managed to lead its competitors out of the mouth of the Chesapeake and into the Atlantic, winning the City of Annapolis trophy in the process, its overall placement at the conclusion of the regatta was sixth, although it did at least finish ahead of the two other American entries. *Chessie Racing*'s organizers no doubt would have preferred a better result, but they still felt they had won the race in deeper ways. They were proud of their role in revitalizing sailing as a sport, particularly in the Chesapeake, and by demonstrating respect for, and interest in, the environment via their collaboration with Living Classrooms. As Kerry Fishback, liaison with Living Classrooms, put it, "Yes, *Chessie*, the big green monster from the Chesapeake Bay, won the respect and admiration of everyone she touched and proved winning isn't everything. It's all the other stuff that happens when you're trying. And *Chessie* most definitely won in that respect."[8]

In 2000, Living Classrooms collaborated on a very different Chessie boat project, this time contracting with Exotic Paddle Boats USA to construct a fleet of custom-designed, dragon-shaped paddleboats for Baltimore's Inner Harbor. Based on a

Chessie paddleboat at Baltimore's Inner Harbor, 2006.
COURTESY OF CHRISTINA L. CHEEZUM

sketch provided by the foundation, the company sculpted a series of sea-serpent-shaped bodies for the boats, complete with plesiosaur necks. Locals naturally dubbed the dragons "Chessie," and they became extremely popular among visitors. "Now Baltimore's Inner Harbor is the home of many Chessies," the company's website reported at the time, "and the location of the annual Chessie paddle boat race each spring."[9]

The paddleboat race is a good example of Chessie's continuing ability to resonate with different aspects of Maryland and Chesapeake identity. The race was introduced as part of Baltimore's Preakness celebrations in May 2001. The second tourney in the Triple Crown horserace, the Preakness Stakes is a Maryland tradition connected to its equestrian past. By climbing in a paddle Chessie and pedaling the Preakness Sea Monster Stakes, children not only joined the broader festivities surrounding the race, they merged another Maryland historical tradition with the Chesapeakiana that Chessie embodied.[10]

The sea monster race also, of course, inevitably celebrates the triumph of recreation over work as the purpose of the waterfront. Hardly a century before the paddleboats dipped their fiberglass scales in the water, the Inner Harbor had been a working seaport. After World War II, when heavy industry began deserting American cities, the Baltimore waterfront floundered and fell into decay. Redevelopment for tourism started in the 1960s but profoundly accelerated in the 1970s and 1980s under the leadership of Baltimore mayor William Donald Schaefer. Schaefer's efforts created the modern waterfront that *Chessie Racing* visited and the paddleboats inhabited, and, going back a few years, that Paul Foer's second Chessie expedition sailed out of in 1987. In a place that was once reserved for watermen and longshoremen, tourists could now spend the day exploring shops, restaurants, and specialized attractions like the National Aquarium and the Maryland Science Center. And they could cap off their Inner Harbor experience by paddling leisurely across it in a Chessie, communing with nature through a sea monster, in an environment refashioned around consumerism and recreation. Even an urban landscape, it turns out, can provide habitat for a sea monster.

If there were any doubt that the Chessie paddleboats provide a vehicle for tourists to satisfy a deep spiritual need for the Chesapeake and its water, consider that when they were first introduced, they were in constant demand, and they still are today. In 2003, the dock manager of the Living Classrooms–operated National Historic Seaport of Baltimore, Richard Slingluff, told a Baltimore newspaper, "On a regular basis, we have to discourage people to ride the dragon. We've had people want to wait for three hours. But . . . we have to discourage rides on the Chessie for safety issues on the dock."[11] The paddleboats' appeal has even caught the imaginations of tourism boosters in other venues around the bay. One official from the Dorchester County Chamber of Commerce, on the Eastern Shore, suggested the Chessies were "very, very easy to notice and a delightful kind of thing" and hoped to import something similar to the newly redeveloping waterfront at Cambridge, Maryland. Back in its heyday, Cambridge, like Baltimore, used to house a thriving seafood industry, a river- and bayfront lined with packinghouses, and no shortage of watermen and shoreline laborers to support it all. With industry in decline and gentrification on the rise, the city now looked to urban renewal efforts to attract tourists and regenerate its ailing economy. Could Chessie paddleboats provide the necessary allure? The chamber of commerce sure thought so. "We are in a community here that is on the verge of being discovered," the same official wrote. "I just felt that the kind of thing that they manufacture [at Exotic Paddle Boats USA] could be worked into a project that would end up being a mark of distinction for the city of Cambridge."[12] To date, Cambridge has never installed its own fleet of Chessies, but the monster is still there in spirit: in July 2009, the Choptank River Fishing Pier, located right outside Cambridge Harbor, was renamed in honor of Bill Burton. In October 2013, Dorchester County adopted a new tourism slogan, "Water moves us." As its promotional website explains, the brand "embraces dorchester's [*sic*] most recognized and distinguishing physical characteristic—water."[13] As ever, though Chessie might not be immediately visible, it remains present, lurking under the surface.

Modern Fictions

The spirit of Chessie has become an anchor by which a region, fascinated by the Chesapeake Bay, can make sense of its cultural and historical complexity. Local authors have shown no desire to let go of the sea monster and embrace the manatee, and no wonder. Charming as the latter animal might be, it is too *real* to be especially useful as a metaphor for anything other than environmental matters. Chessie the sea monster, on the other hand, speaks to a whole range of issues: cultural, economic, environmental, historical, and more. Even children's authors have gotten more mileage out of the monster than the manatee.

In his 2000 black comedy, *Key Monster*, Lee Dravis, like John Barth before him, employs Chessie as a catalyst for a pivotal decision that the main character, Eugene "Winch" Winchell, makes in the course of the novel. Winch is a former watermen who grew up on Kent Island; even his nickname alludes to the winches used on fishing boats. After his father, also a "lifelong waterman," dies at the age of 62 while oystering, and his mother's death from cancer not long after, Winch sinks into substance abuse and shiftlessness. With little to hold him on the island, and perhaps some resentment toward the waterman's life for leading to his father's untimely death, Winch plots an escape from his problems. He buys a cabin cruiser, which he dubs the *Stuffed Ham*, at an IRS auction and prepares to sail it around the world. His plans are dashed, however, when a hurricane destroys his family dock and the *Stuffed Ham* is mildly damaged. In debt for repairs and now renting a slip, Winch turns to tourism, "offering luxurious cruises-to-nowhere aboard the *Stuffed Ham*," in order to make ends meet.[14]

The moment of truth in Winch's life is seeing Chessie. Is the thing he sees breaking the surface a manatee, as in the mid-1990s, or just a drug-induced hallucination? Winch has a classic encounter: "The horrible thing swiveled its serpentine neck and gazed placidly back at him, lidless black eyes filling the circular field of Winch's binoculars. The head rose up, almost even with the boat's flying bridge and Winch was forced to bend backwards to keep Chessie's face in sight." But Winch is a waterman and watermen do not believe in sea monsters: "He could tell no one he'd seen the legendary monster of the Chesapeake, not if he wanted to hang onto his last tattered shreds of dignity on the island. His credibility among the other watermen was already approaching zero. Babbling about seeing a sea monster was the last thing he needed."[15] Estranged from his home, running out of money, chased by a corrupt investment banker wanting the *Stuffed Ham* back, and fearing he is crazy, Winch cuts his losses, sells his waterfront family home, and departs for new waters in Key West.

Despite the broadness of its plot, *Key Monster* nonetheless reflects some of the real problems facing watermen and other blue-collar people living around the Chesapeake. The comedy of errors that drives Winch to Key West might be improbable, but the forced departure was no joke, and definitely no illusion. As we have seen, as early as the 1980s, public figures on Kent Island were predicting that low-income residents whose families had lived there for generations would not be able to afford to stay, due to the explosion of high-dollar residential development. Winch's childhood home and family dock are ultimately destroyed by a freak hurricane, but in real life they would have been cleared to make way for a condominium complex—a process that, by the time Dravis was writing, was no less a force of nature than if it *had been* a destructive act of God.

It is especially poignant that it is a Chessie encounter that convinces Winch to turn his back on Kent Island and the Chesapeake. We learn that despite his own challenges, Winch's father tried to retire but "couldn't seem to leave the bay behind," and Winch seems to be made of similar stuff.[16] Why does a brush with Chessie shake Winch's worldview so profoundly? Perhaps it is the uncanny, impassive way Chessie stares down Winch during the pivotal encounter. Maybe Dravis was just exceptionally visionary: in 2003, the Maryland Watermen's Monument was erected at the Kent Narrows. Although the statue, depicting two watermen working from their boat, was designed to memorialize the industry's unique culture and legacy, it is difficult not to see it instead as an acknowledgment of the industry's *end*. If the monument were situated a little closer to the water than it is, it would be easy to imagine Chessie swimming past on its endless travels, stopping only to gaze unemotionally with lidless eyes at the watermen ossified on their plinth, before proceeding, inexorably, on its way. That was the way of this elemental force of a monster: transformative but aloof and impassive—especially so when watermen were involved.

Yet, Chessie does not represent only endings in *Key Monster*. The creature appears periodically throughout the book in a framing device depicting its migration out of the bay alongside Winch. Dravis puts the reader inside Chessie's mind, and we discover that its overriding goal is to fulfill its life cycle: to eat, mate, and then ultimately die. The monster's story is thus a metaphor for rebirth and renewal, forces Winch himself experiences in the course of the novel. The use of Chessie as a fertility symbol is not original to Dravis. Barth had employed the monster in the same way, although perhaps more subtly, in his linked novels about sailing on the Chesapeake: *Sabbatical: A Romance* (1982) and *The Tidewater Tales* (1987). In both of those works, characters grapple with issues surrounding pregnancy, progeny, and legacy. As Dravis does in *Key Monster*, Barth uses Chessie to symbolize the moment pivotal decisions are made, but Barth takes the metaphor one step further. In his typically earthy way, he imagines the sea serpent as a sperm cell travelling into the womb of the Chesapeake Bay, leading ultimately to the conception of something new—whatever it might be. This Chessie is a life force, an expression of nature's most basic imperatives.

It is much more wholesome than Barth's or Dravis's adult works, but Lisa Cole's 2011 juvenile novel, *Searching for Chessie of the Chesapeake Bay*, explores similar themes. Scheduled to attend a summer kayaking camp near Kent Island, the book's protagonist, 10-year-old Claire Meadows, becomes fixated on finding Chessie. First, she has to convince her skeptical older brother, Jim, to help her, and the bulk of the book explores their strained relationship both with each other and with their mother. Their father is not present in the book, and his absence and the shadow it casts over the family are mirrored by the mystery of Chessie. Scary and reptilian, with "dagger-like teeth" and "fierce-looking" eyes, this Chessie is aloof and alien like the Dravis and Barth monsters, and not much like the friendly creature real eyewitnesses described.[17]

Chesapeakiana is at the book's core. "From the first time Lisa Cole set sail in the Chesapeake Bay she knew there was more to the bay than a unique sailing experience," the *Children's Book Review* declares in its online description of the book. "She spent the next three years researching, sailing, and kayaking in and along the bay's waters. During that time she kept a journal of her encounters with marine life, storms, and the communities lining the shores."[18] Accordingly, Cole weaves in a family friend, Captain Jack, who was once a Chesapeake Bay waterman whose "family prospered there for generations."[19] The Bay Bridge also makes an appearance, "looming" over "choppy waves" and "vast shoreline," and when the children cross it they are treated to the full spectrum of watercraft, from naval vessels to workboats, and everything in between. Claire, especially, yearns for the possibilities of the bay, although it is her brother who voices the mystique evoked by the monster. Contemplating the difficulty of finding Chessie during rough weather conditions on the bay, he muses to himself that "everything is connected in nature."[20]

In contrast to this romanticism, Cole presents a Kent Island that has had all of its naturalness plated over. When the family arrives there, for instance, the children are confronted, and depressed, by suburban development. "This doesn't look like a wilderness to me," Jim grumbles at one point. "How many shopping centers and houses can there be?" He goes on to complain about the "cookie cutter" nature of the sprawl.[21] As we have seen, it is a familiar refrain in the story of Chessie, but it is unusual to see it make an appearance in any fictional treatment of the monster, let alone in a children's book.

Unsurprisingly, Cole uses Chessie as a proxy to discuss environmentalism, although, reminiscent of Dravis and Barth, the monster does so as an expression of nature's life force. "Fixing the problems of the bay is a gigantic task," one character tells Claire, showing the girl all the trash littering the sea floor. "Humans are forgetting what's important. Life will be broken without the bay. Birds and bay creatures will disappear. Water will no longer be the giver of life without the bay grass and the oysters to purify it."[22] Chessie's role in this crisis, Cole seems to say, is to rescue the bay in the same way that it saves Claire's life at one point—a job for which the monster is uniquely qualified, since it embodies the natural and cultural mystique of the bay. "Mom, it's so beautiful in the bay's waters—there is so much life," Claire exclaims at the end of the book. "The sea creatures live together in harmony, where one needs the other to survive. . . . And Chessie helps them all."[23] It seems unlikely that even Glenn Kinser and his staff could have summed up Chessie's significance for the bay more succinctly than this.

A Vernacular Feel

That Chessie has become one of the most recognizable extensions of the Chesapeake was demonstrated again, in grander fashion than ever, when a new attraction arrived at Baltimore's Inner Harbor in 2012. The venue was a Ripley's Believe It or Not! museum (or "odditorium," to use the term the Ripley's people prefer)—probably the most touristy of tourist traps. Chessie's contribution to the enterprise was in the form of a massive fiberglass sculpture, painted a brilliant green, coiled around the second-floor entranceway of one of two waterfront pavilions at the Harborplace shopping center. Patrons could scarcely miss the message: Maryland might harbor a museum full of oddities, but Chessie remained its star. "We like to have a local vernacular feel to everything we do, and Chessie is very much that, very much a myth of the area," a Ripley's spokesman told the papers a few months before the museum opened its doors.[24]

Local or not, this iteration of Chessie almost never made it to the Inner Harbor. Public response was lukewarm when the museum was first mooted in the fall of 2011, and the first design for the facade drew especial disapproval. Initially, Ripley's imagined Chessie as a ferocious dragon straight from a medieval *mappa mundi*, a "three-dimensional sea monster bursting from the building, teeth bared, as its green body coiled around a three-masted ship," as the *Baltimore Sun* reported it. City planning officials baulked at the proposal because they feared its size and gaudiness would establish a precedent that would disrupt the existing design aesthetic in and around Harborplace. They also found the representation of Chessie to be too "fierce-like"—an apt criticism given the serpent's reputation for friendliness, although the concern was probably more that the building itself was welcoming rather than that the monster was correctly characterized.[25]

Interviewed by the *Baltimore Sun*, Laurie Schwartz, executive director of the Waterfront Partnership, a booster organization for the Inner Harbor, called the inclusion of Chessie "playful" and a "fun component," although she worried that the entrance as first designed would "overpower the harbor." Others were less charitable, even toward the monster. Klaus Philipsen, of the Baltimore chapter of the American Institute of Architects, found Chessie "rather crass" and an example of "a tendency to turn the Inner Harbor into a carnival." For some of the *Sun*'s interviewees, the proposition of a life-size Chessie at the harbor evoked a similar installation on the boardwalk in Ocean City, Maryland, where a shark erupted out of the second story of a Ripley's storefront. It was not a congenial comparison: the prevailing opinion was that the Inner Harbor's waterfront demanded greater dignity and good taste than was on display at the beach resort. Philipsen, in particular, stressed the importance of maintaining a design aesthetic consistent with the existing pavilions. "They were conceived in a nautical way," he explained, "with a good view of the water from all places, and they were outward-oriented."[26] Philipsen said more than he realized: divorced from subjective considerations of taste, his words might equally have described the Chessie phenomenon itself. And in any event, was a giant fiberglass sea monster sculpture really a more vulgar imposition on the Baltimore waterfront—a space once devoted to industry—than a shopping center specifically designed to allow customers to admire the water?

Ripley's, no doubt well acquainted with accusations of bad taste, dutifully answered the public's concerns about the entranceway, toning and *shrinking* down its menacing Chessie and replacing the original sailing ship with a dinghy. Visitors could climb into the boat and, in a reversal of typical sea monster sighting procedure, pose for pictures of *themselves*. The new design evidently satisfied Baltimore's planning department, as renovations to the building commenced in the spring of 2012 and the museum opened its doors on June 9. With two floors, totaling 12,500 square feet and 500 displays, the odditorium made for an unorthodox but no less flashy addition to Harborplace's collection of chain restaurants and shopping venues, many of which debuted at the same time, but with much less controversy. Alas, however, Chessie's twenty-first-century sojourn in the Inner Harbor lasted less than a decade. Like other businesses in the tourism industry, the odditorium closed its doors in early 2020 due to the COVID-19 pandemic. Ripley's made the closure permanent in May 2020, dismantling the museum, along with the Chessie sculpture, in a few weeks. It was an abrupt disappearance, very much in keeping with the monster's habits of old.

The obvious distaste the Baltimore elite had for Ripley's notwithstanding, the company was merely doing with Chessie in a big way what people living near the Chesapeake had been doing since the 1980s: capitalizing on their local oddity. In the last 40 years, the Chessie of popular imaginings has become a cottage industry across the Chesapeake, appearing in the public square in ways that (usually) remind the observer of the region's peculiar milieu and charms. Boat dealer Chessie Marine Sales, in Elkton, Maryland, is an obvious one, boasting not only the monster's name but also a striking logo featuring a plesiosaur. A slightly more obscure connection to the bay occurs when one visits RAR Brewing in Cambridge, Maryland, and elects to have a Chessie Burger for dinner. The brewing company's restaurant, where you can buy said burger, is *also* called Chessie Burger, and its website and logo feature another charming iteration of a plesiosaur-style Chessie bobbing in rippling waters. The venue makes clear that it is located "on the shores of the Chesapeake Bay."[27]

Western Shore residents have also continued to embrace Chessie. In October 2014, a couple from Crofton, Maryland, produced a children's music festival, called Chessie Jam, at the Baltimore Inner Harbor. Designed to be "something hip, something different from your average kids show," the festival came complete with a costumed mascot not unlike the one the US Fish and Wildlife Service had once used.[28] For the older crowd, epicures in Baltimore today can visit the Union Craft Brewery and imbibe some Barrel-Aged Chessie ale. According to their website, "emerging from the depths of the Chesapeake Bay, this rarely experienced brew" was introduced—"surfaced," as they put it—to celebrate the anniversary of the brewery's founding.[29] Down in Annapolis, if your boat needs some work or you need a diver, you can avail yourself of Chessie Marine Services. Their

Logo for RAR Brewing's Chessie Burger, 2023.
COURTESY OF CHRIS BROHAWN

logo looks to have been adapted from the familiar Fish and Wildlife Service image. Washington, DC, metro area high schoolers who play hockey can do so with a team called the Chesapeake Bay Monsters (a.k.a. Chessie Hockey); they even play the occasional game at Kent Island. Youngsters near Alexandria, Virginia, can spend the day at Chessie's Big Backyard, a playground and Chesapeake-themed nature trail. The monster itself—in sculpture form—even makes an appearance at the park's sprayground. Doubtless there are many, many more manifestations of the monster at local events, on signage, in public artwork, and on the printed page all over the Chesapeake. Even if the "real" Chessie the sea serpent is never sighted again, such ubiquity will assure the creature a place in the public's imagination for a very long time to come.

Chessie Forever

Back in the summer of 1993, during the interregnum between monster sightings and manatee madness, when it looked like the Chessie craze had finally burnt itself out, Baltimore *City Paper* reporter David Dudley interviewed Bill Burton for a feature exploring why the monster had seemingly vanished. Burton, whom Dudley portrayed as a cross between the Ancient Mariner and Robert Shaw's grizzled sailor, Quint, in *Jaws*, seemed to blame the monster's disappearance on overexposure, the result of the "circus atmosphere" the creature had inspired since the first sightings. Desiring to right this wrong, Burton thought about publishing his own account of the monster's story, but not "the usual tabloid monster hunt," Dudley wrote. Burton's book would be "a story about people. And what they saw out on the water. Nothing more."[30]

Burton's aversion to the "circus" surrounding Chessie and his proposed "just the facts" book on the creature were a critique of the press rather than the general public. "Something can be so popular no one believes it," Burton complained. This was not an expression of Menckenian elitism, a swipe at the credulous masses. The media, he felt, had fumbled the Chessie story by reporting it in a sensationalist way. It was not so much that they made sport of Chessie, but that in indulging their fantasies of a "sea monster" they had numbed the public to the *actual* phenomenon that *genuinely* existed and demanded scientific investigation. Chessie needed to be treated seriously, and the press had followed the path of least resistance and embraced the "circus" instead. It was a view that probably most of the monster's boosters shared, and while it was understandable, it was also wrongheaded.

The prospect of a "just-the-facts" book about Chessie, stripped of all the hoopla, is a noble goal, but an impossible one. To begin with, no one really knows what people *actually* saw on the water, and what they reported was infuriatingly variable, both in content and reliability. Eyewitness accounts were often not very detailed and either diverted wildly from the "classic" serpent description or replicated it so slavishly that they smacked of perception bias. The narrative Burton imagined, then, would have made for either incredibly frustrating reading, or a fanciful—and ultimately inaccurate—attempt to force the available evidence into a coherent whole. The logical end to it would have been one of two possibilities: that the Chesapeake Bay was seemingly infested with dozens of minutely varying species of cryptids, or that a single, monolithic serpent or colony of serpents was virtually ubiquitous in the bay and its tributaries—in either case, after having been seen only rarely, if not never, before 1978. Both propositions were silly and ignored a simple truth that should have given pause to an accomplished newsman like Burton: that multiple eyewitness accounts of any event, let alone a sea serpent encounter, are notoriously difficult to harmonize. Taken by themselves, the accounts are the *least* revealing part of the Chessie story and amount to less than the sum of their parts. There *had* to be something more to Chessie than those.

What Burton missed about the whole Chessie saga was that the sensationalism he found so distasteful was integral to the creature's existence. The monster and the hoopla existed symbiotically because both were extensions of the microscopic obsession with every aspect of the Chesapeake Bay—economic, environmental, scientific, political, cultural, developmental—that gripped the region so fiercely after World War II. Taking issue with the circus atmosphere surrounding Chessie, let alone trying to prove the creature really existed, was an act as futile as King Canute trying to hold back the tide.

There is no way to disentangle the rise of the monster from the rise of Chesapeakiana. The two have always been, and will forever be, linked. For cryptozoologists, even accidental ones like Burton, that link was an impediment, the circus that undermined serious investigation of the creature. For environmentalists, it was a boon that enabled them to reach and educate a vital new constituency. For tourists and suburbanites, the link enabled them to commune with and take ownership of the bay and its waterfront as they sought out their own slice of the American dream. For watermen and natives, Chessie's attachment to the bay was an annoyance at best, but one that trivialized their livelihood and augured the decline of their economic and social authority.

Today, Chessie persists in the public consciousness and in local lore, particularly in Maryland, despite a lack of fresh sightings. On National Sea Serpent Day, August 7 of each year, the monster usually receives a nod of recognition, and it still pops up in the occasional documentary or retrospective about weird phenomena around the state and nation. Even Chessie the manatee continues to appear, defying scientists' periodic pronouncements that he has died. As late as January 2022, he was discovered lounging in warm waters near Fort Lauderdale.[31] Ultimately, improbably even, Chessie has turned out to be the Chesapeake Bay's most durable ambassador. Perhaps that is the measure of a truly successful monster: that its mythology keeps it alive even when rational thought insists it should die off.

But even if sightings never come in again, or the manatee's radio transmitter fails for the final time, Chessie the friendly sea serpent will still live on. Children living around the bay will still find the idea of a cryptid in their backyard enthralling, and Chessie will *never* lose its environmental cachet. Today, practically anything related to the bay has the potential to evoke Chessie: tourism projects that sell Chesapeake heritage, for instance, or ongoing plans to build a third span of the Bay Bridge, or, maybe most tellingly, the recent effort to have the Chesapeake Bay designated as a National Recreation Area by the National Park Service.[32] Chessie is everywhere: a companion animal to two generations of people presently living along the bay's shoreline and the many more who inevitably will follow them.

Although Chessie was born and grew up in a cynical age, and sometimes reflected the worst excesses of postwar consumerism and suburban sprawl, in the end it evolved beyond those origins to become something truly iconic. Whatever his feelings about the sea serpent's notoriety, even Burton yielded to that central fact of its existence. "You want [Chessie] to be a symbol," he told Dudley. "A symbol for what the bay is all about."[33] Propelled by such powerful belief, this bay monster will keep making waves forever.

ACKNOWLEDGMENTS

By the time this book makes it to shelves, it will be a project 20 years in the making. As a consequence, I have accrued many debts for help along the way. Much of the early research for this book was conducted for my dissertation, and I wish to renew my thanks to the many contacts who opened doors to me when I was first charting Chessie's course. These include Patty Goddard, Larry Hall, Marna Hostetler, Glenn and Lillian Kinser, Mark Madison, Jeffrey Marshall, Min Park, and Cecilia Petro. I am also grateful to the many interviewees who shared their memories of Chessie with me, and to my loyal transcriptionists, Christy Cheezum and Dan Tortora, whose efforts made them available for research.

I owe a debt to another round of helpers during the preparation of the present book. Ann Reinecke and the Caroline County Public Library supplied me with numerous interlibrary loan requests, no matter how obscure; and Becky Riti and the Talbot County Free Library provided access to newspaper collections not presently digitized. Ellen Alers of the Smithsonian Institution Archives helped with research in newly available archival holdings. Thanks are also due to my friends Beth Carmean and Margaret Iovino, and my niece, Susannah Cheezum, for serving as occasional research assistants.

In the area of illustrations, I am thankful to all who gave permission to use their images—particularly Lindsay Leavitt and the Kyker family, and Trudy Guthrie, for the use of their Chessie sketches. I am also grateful to Christy Cheezum, for her help formatting images, and to Erin Greb, who prepared maps for me.

The book has also benefited from the efforts of numerous friends and colleagues who reviewed and proofread it, including Paul Lewis, Susan McCandless, Linda Pevey, and Derek Simmons. At Johns Hopkins University Press, editors Laura Davulis and Ezra Rodriguez made the publishing process smooth and welcoming, and Robert E. Bartholomew and an anonymous reviewer gave useful advice that improved the work immeasurably. Michelle Scott and her team at Westchester Publishing Services rescued me from many errors typographical and conceptual. Any defects of fact or interpretation are my responsibility alone.

To fund the publication of this book, I received a very generous grant from the Peck Family Foundation. I cannot express sufficiently my gratitude for the Peck family's support of my scholarship over the years. Likewise, this enterprise would not have come to fruition without the help and encouragement of my own family. They have been on this expedition with me for a long time, and I hope they appreciate the final product. My brother, Tom, and I grew up with Chessie. Little did he know, when he wrote a letter about the monster to the local newspaper back in August 1984, that it would resurface, Chessie-like, 40 years later!

Finally, for Dan Tortora I reserve especial gratitude. Across many years and life changes for both of us, he pushed me to complete this book and gave freely of his time and talent to see that I got it done. I cannot thank him enough for his persistence and encouragement, and most of all for his friendship.

NOTES

In the notes, the following abbreviations have been used for interviews, manuscript collections, and frequently cited newspapers.

Interviews

Transcripts of all interviews are in the author's possession.

Bangert Interview	Kathi Bangert, telephone interview by author, July 27, 2005.
Della Interview	George Della, telephone interview by author, June 14, 2005.
Foer Interview	Paul Foer, telephone interview by author, August 12, 2008.
Frizzell Interview I	Michael Frizzell, telephone interview by author, September 24, 2005.
Frizzell Interview II	Michael Frizzell, telephone interview by author, October 8, 2005.
Guthrie Interview	Gertrude T. Guthrie, telephone interview by author, April 22, 2023.
Holland Interview	Jeffrey Holland, telephone interview by author, July 20, 2005.
Kinser Interview	Glenn Kinser, telephone interview by author, March 21, 2005.
Slattery Interview	Britt Slattery, telephone interview by author, May 25, 2005.

Manuscript Collections

GKPA	Glenn Kinser Personal Archive. Copies in author's possession.
PFPA	Paul Foer Personal Archive. Copies in author's possession.
SIA 12-057	National Museum of Natural History, Department of Vertebrate Zoology, Curatorial Record, Smithsonian Institution Archives, SIA Acc. 12-057.
SIA 95-031	International Society of Cryptozoology, Records, 1981–1991, Smithsonian Institution Archives, SIA Acc. 95-031.
SJR13-Della	Documents relating to Senate Joint Resolution 13 (1985), received from the office of Senator George Della. Copies in author's possession.
SJR13-MD	Documents relating to Senate Joint Resolution 13 (1985), received from Maryland Department of Legislative Services. Copies in author's possession.

Frequently Cited Newspapers

AC	*Capital* (Annapolis, MD)
BES	*Evening Sun* (Baltimore, MD)
BS	*Baltimore Sun*
KIBT	*Kent Island Bay Times* (Chester, MD)
NYT	*New York Times*
RTD	*Richmond (VA) Times-Dispatch*
SD	*Star-Democrat* (Easton, MD)

Introduction

1. On these points, see Paul Semonin, *American Monster: How the Nation's First Prehistoric Creature Became a Symbol of National Identity* (New York: New York University Press, 2000).
2. On the history of monsters, sea monsters, sea serpents, and their study, see Brian Regal and Frank J. Esposito, *The Secret History of the Jersey Devil: How Quakers, Hucksters, and Benjamin Franklin Created a Monster* (Baltimore: Johns Hopkins University Press, 2018), 59–66, 92–97.

3. Chandos Michael Brown, "A Natural History of the Gloucester Sea Serpent: Knowledge, Power, and the Culture of Science in Antebellum America," *American Quarterly* 42, no. 3 (September 1990): 402–36; William Elliott, *Carolina Sports by Land and Water: Including Incidents of Devil-Fishing, Wildcat, Deer, and Bear Hunting, Etc.* (Columbia: University of South Carolina Press, 1994), 95–109; Argus, "Too Much Lenity," *Morning Herald* (St. Joseph, MO), February 18, 1862.
4. A. C. Oudemans, *The Great Sea-Serpent: An Historical and Critical Treatise* (Leiden: E. J. Brill, 1892). Oudemans's conclusions can be found on pages 546–72.
5. "Sea Serpent from Kansas," *Daily News and Intelligencer* (Mexico, MO), December 31, 1900; "The Sea Serpent Is a Fact," *Washington Times*, April 24, 1904.
6. "Useful in His Day," *Wood County Reporter* (Grand Rapids, WI), October 7, 1915; "The Sea Serpent," *Grand Rapids Press*, quoted in *BES*, July 15, 1915. On U-boats, see, for instance, E.E.L., "The German Incarnation of a Sailors' Myth," *BS*, March 8, 1917.
7. *BES*, August 19, 1921; *US Navy Review* quoted in *Miami Daily News*, May 6, 1934.
8. See, for instance, Robert J. Menzies, "I Fished for a Sea Serpent," *This Week*, January 29, 1961, 16–17.
9. Don Shoemaker, "Cartership Down," *Miami Herald*, September 12, 1979. On Carter's scrape with the "killer rabbit," see David Farber, "The Torch Had Fallen," in *America in the Seventies*, ed. Beth Bailey and David Farber (Lawrence: University Press of Kansas, 2004), 19–20.
10. The standard work on the decade of the 1970s is Bruce Schulman, *The Seventies: The Great Shift in American Culture, Society, and Politics* (New York: Free Press, 2001). On the "crisis of confidence" and other historical themes of the decade, see Bailey and Farber, *America in the Seventies*, esp. 1–49, 157–80.
11. For a comprehensive bibliography of cryptozoological works that reaches back to the origins of the genre, see Karl P. N. Shuker, "A Bibliography of Cryptozoological and Zoomythological Books," http://www.karlshuker.com/bibliography.htm.
12. Paul Cronin, *Werner Herzog—A Guide for the Perplexed: Conversations with Paul Cronin* (New York: Farrar, Straus and Giroux, 2014), 331; David J. Halperin, *Intimate Alien: The Hidden Story of the UFO* (Stanford, CA: Stanford University Press, 2020), 46 (emphasis in original).

13. US Department of Commerce, Bureau of the Census, *1980 Census of Population*, vol. 1, *Characteristics of the Population*, chap. B, *General Population Characteristics*, pt. 48, *Virginia* (Washington, DC: US Department of Commerce, Bureau of the Census, 1982), 11–12; US Department of Commerce, Bureau of the Census, *1980 Census of Population*, vol. 1, *Characteristics of the Population*, chap. B, *General Population Characteristics*, pt. 22, *Maryland* (Washington, DC: US Department of Commerce, Bureau of the Census, 1982), 11.
14. The classic study on the subject is Kenneth T. Jackson, *Crabgrass Frontier: The Suburbanization of the United States* (New York: Oxford University Press, 1985), which provides both an overview of the historical mechanics of the rise of suburbanization in America and some rather naïve predictions about its potential longevity. The role of affluence in the period is explored in Lizabeth Cohen, *A Consumers' Republic: The Politics of Mass Consumption in Postwar America* (New York: Knopf, 2003).
15. Kyle Riismandel, *Neighborhood of Fear: The Suburban Crisis in American Culture, 1975–2001* (Baltimore: Johns Hopkins University Press, 2020), 3.
16. Historians have not devoted much attention to the cultural fallout from postwar suburbanization in rural areas, perhaps because American historians tend to write from an urban or suburban perspective. Although it is primarily about the history of tourism, Hal K. Rothman's *Devil's Bargains: Tourism in the Twentieth-Century American West* (Lawrence: University Press of Kansas, 1998) delivers a penetrating and acerbic analysis of the process by which outsiders terraform rural communities, as well as the wrenching social and cultural costs locals pay as a result of that process. Rothman's introduction, in which he comments on the "symbolic power" of whales in the waters off the island of Maui, is especially germane to the story of Chessie.
17. On this trend, see Bruce J. Schulman, "The Privatization of Everyday Life: Public Policy, Public Services, and Public Space in the 1980s," in *Living in the Eighties*, ed. Gil Troy and Vincent J. Cannato (New York: Oxford University Press, 2009), 167–80.
18. Historians have found the 1980s more difficult to summarize than other decades. This interpretation, using Reagan as a lens, derives from Gil Troy's historical biography, *Morning in America: How Ronald Reagan Invented the 1980s* (Princeton, NJ: Princeton University Press, 2005).
19. Robert McMorris, "Miamians Find Another Reason to Stay at Home," *Omaha World-Herald*, July 13, 1982.

Chapter 1. Catching Monster Fever

1. Anne Hazard, "Residents Think Creature's Real," *RTD*, August 20, 1978; Anne Hazard, "Sea Creature Is Reported in River," *RTD*, August 15, 1978.
2. Anne Hazard, "More Claim to Have Seen Strange Aquatic Creatures," *RTD*, August 17, 1978.
3. The figure of 20 people who had claimed to see the monster is from Hazard, "Residents Think Creature's Real"; news brief, *Richmond News Leader*, August 16, 1978.
4. Hazard, "Sea Creature Is Reported."
5. Merriner quote: Hazard, "More Claim"; Douglas quote: Anne Hazard, "Like Serpent, Bid to Help It Goes Under," *RTD*, August 19, 1978.
6. Anne Hazard, "Sightings Still Cause Controversy," *RTD*, September 3, 1978.
7. Hazard, "Residents Think Creature's Real."
8. Anne Hazard, "Farmer Says He Saw Serpent 30 Years Ago," *RTD*, August 18, 1978.
9. Chad Arment, "Giant Snake Stories in Maryland," *INFO Journal*, no. 73 (Summer 1995): 15–16.
10. Hazard, "Residents Think Creature's Real."
11. On suburban environmental anxieties and the pervasive NIMBY movement in America, see Kyle Riismandel, *Neighborhood of Fear: The Suburban Crisis in American Culture, 1975–2001* (Baltimore: Johns Hopkins University Press, 2020), 16–45.
12. "Nutria: An Invasive Species," Maryland Department of Natural Resources, https://dnr.maryland.gov/wildlife/Pages/plants_wildlife/invasives/inv_Nutria.aspx. On Virginia's travails with the nutria, see Joanne Kimberlin, "Menace to the Wetlands," *Daily Press* (Newport News, VA), May 30, 2021.
13. Hazard, "Residents Think Creature's Real."
14. Hazard.
15. Anne Hazard, "Two More Report Seeing Serpent," *RTD*, August 25, 1978.
16. Hazard, "Sightings Still Cause Controversy"; the Virginia Beach encounter appears in "4 See Creature in Linkhorn Bay," *RTD*, August 30, 1978.
17. Hazard, "Sightings Still Cause Controversy."
18. "Rational" explanations for the serpent appear in nearly every news article. The submarine suggestion is from Hazard, "Residents Think

Creature's Real"; the drug trafficking suggestion is from Hazard, "Sightings Still Cause Controversy."

19. Hazard, "Like Serpent."
20. Tom Howard, "Hoot, Mon: Leave the Scots' Nessie Alone We Have Chessie Herself Closer to Home," *RTD*, February 9, 1994.
21. See, for instance, "'Sea Serpent' Sighted," *Syracuse (NY) Herald-Journal*, August 18, 1978; and "Chesapeake Has Monster Too," *Chillicothe (MO) Constitution-Tribune*, August 18, 1978.
22. Syd Courson, "Potomac Sea Serpent Grants Exclusive (Mythical) Interview," *News* (Frederick, MD), September 1, 1978.
23. Richard D. Lyons, "A 'Monster' Sighted in Chesapeake Bay," *NYT*, November 19, 1978.
24. William Worthington, letter to the editor, *NYT*, December 3, 1978. "Daniel Webster and the Sea Serpent" was originally published in Stephen Vincent Benét's short story collection *Thirteen O'Clock: Stories of Several Worlds* (New York: Farrar and Rinehart, 1937).
25. Anne Hazard, "'Monster Fever' Has Cooled," *RTD*, January 8, 1979; Brad Cavedo, "Memories Strong of '78's 'Chessie,'" *RTD*, November 4, 1979.
26. Brad Cavedo, "'It Looked like a Snake,' Potomac Farmer Says," *RTD*, June 20, 1980.
27. "Return of 'Chessie,' the Mysterious Sea Creature of the Potomac," *NYT*, June 22, 1980.
28. Albert Oetgen, "Non-monster 'Very Explainable,' Latest Chessie Observers Say," *RTD*, September 21, 1980.
29. Brad Cavedo, "Second Sighting Reported," *RTD*, June 25, 1980.
30. Cavedo.
31. Charles McDowell, "The Sea Monster in the Potomac," *RTD*, July 1, 1980.
32. Hazard, "'Monster Fever' Has Cooled."
33. Brad Cavedo, "Life Altered for 'Chessie' Sighters," *RTD*, June 29, 1980.
34. Brad Cavedo, "Chessie? Serpent Searchers Insist Creature Lurks Deep Below," *RTD*, August 10, 1980.
35. The Kyker quote is from Hazard, "Residents Think Creature's Real"; the Lawson quote is from Hazard, "Sightings Still Cause Controversy."
36. Albert Oetgen, "Chessie Sighter Comes Forward," *RTD*, November 27, 1980.
37. "Snap Proof, Dalton Advises," *RTD*, July 3, 1980.
38. "'Chessie' Picture Would Bring $50," *RTD*, July 8, 1980; "Sum Doubled for Picture," *RTD*, July 22, 1980.

Chapter 2. A Change of Scene

1. Brad Cavedo, "Chessie? Serpent Searchers Insist Creature Lurks Deep Below," *RTD*, August 10, 1980.
2. Albert Oetgen, "Non-monster 'Very Explainable,' Latest Chessie Observers Say," *RTD*, September 21, 1980.
3. Oetgen.
4. Guthrie Interview.
5. Guthrie Interview.
6. Oetgen, "Non-monster 'Very Explainable.'"
7. Bill Burton, "'Creature' Is Said to Rise from Bay," *BES*, September 17, 1980; Mary K. Tilghman, "Couple Reports Seeing Creature," *SD*, September 16, 1980.
8. AP, "Unidentified Swimming Objects Thrive in Bay," *AC*, September 18, 1980.
9. Burton, "'Creature' Is Said"; Oetgen, "Non-monster 'Very Explainable.'"
10. Anne Hazard, "More Claim to Have Seen Strange Aquatic Creatures," *RTD*, August 17, 1978; Anne Hazard, "Like Serpent, Bid to Help It Goes Under," *RTD*, August 19, 1978.
11. Bill Burton, "Monster Sightings Raise Questions," *BES*, October 17, 1980.
12. Bill Burton, "The Creature from Eastern Bay," *Chesapeake Bay Magazine*, November 1980, 41.
13. Burton, "Monster Sightings Raise Questions."
14. Bill Burton, "Sea Creature Seen in the Gunpowder," *BES*, November 3, 1980.
15. Bill Burton, "Annual Chesapeake Boat Show Features One Daring Trans-Atlantic 'Rag-Mopper,'" *BES*, January 16, 1981.
16. Burton.
17. Charles H. Kepner, letter to the editor, *SD*, September 18, 1980; Mary K. Tilghman, "Monster: Sea Creature May Have Been Harbour Seal," *SD*, September 19, 1980.
18. Burton, "Creature from Eastern Bay," 40.
19. Albert Oetgen, "Fishermen Report Seeing Chessie," *RTD*, October 9, 1980.
20. Oetgen.
21. Albert Oetgen, "'Chessie' Sighter Comes Forward," *RTD*, November 27, 1980; Oetgen, "Non-monster 'Very Explainable.'"

22. Guthrie Interview.
23. Burton, "Creature from Eastern Bay," 41. The term appears in many other Burton articles as well.
24. Candus Thomson, "Bill Burton," *BS*, August 11, 2009.
25. James H. Bready, "Let's Have a Look at You, Essie," *BES*, November 26, 1980.
26. Bready.
27. Albert Oetgen, "'Chessie' Weather Creeps In," *RTD*, April 10, 1981.
28. Albert Oetgen, "What's Happened to Elusive 'Chessie'?," *RTD*, December 27, 1981.
29. Charlotte Rosier is quoted in Russ Robinson, "Chessie May Have Made Video Debut," *BS*, July 11, 1982; Steve Rosier is quoted in Mary Siemer, "Is There a Chessie? Sightings Reported," *SD*, June 30, 1982. Other details of the sighting are drawn from these articles, as well as from Mary Siemer, "'Chessie' Sighting Claimed; Serpent Said Caught on Film," *KIBT*, June 30, 1982.
30. Comments on the film come from Mary Siemer, "'Chessie' Sighting Claimed; Serpent Said Caught on Film," *KIBT*, June 30, 1982; the Frew quote and other details are from Russ Robinson, "Chessie May Have Made Video Debut," *BS*, July 11, 1982.
31. "More Bark than Bite: Appetite for Monsters," *NYT*, July 18, 1982.
32. For the period of 17 years, see, for instance, Siemer, "'Chessie' Sighting Claimed."
33. The price quoted here comes from an advertisement for Easton Firestone, *SD*, October 29, 1982. Determining the average price of a video camera in Maryland in the early 1980s is difficult. Before the advent of the combined camera and recorder unit that we now generically call the "camcorder," which seems to have happened around 1982, the two devices had to be purchased separately. A survey of ads in the Annapolis *Capital* between 1981 and 1983 suggests that the two together would have cost consumers about $2,000 (in 1982 dollars). Frew seems to have used a combined unit to make his film. Income data are derived from US Census Bureau, *Historical Income Tables: Counties*, "Table C1. Median Household Income by County: 1969, 1979, 1989 and 1999," https://www2.census.gov/programs-surveys/decennial/tables/time-series/historical-income-counties/county1.csv; and adjusted to 1982 values using "Seven Ways to Compute the Relative Value of a U.S. Dollar Amount—1790 to Present," MeasuringWorth.com, https://www.measuringworth.com/calculators/uscompare/.

34. John Barth, *Sabbatical: A Romance* (1982; repr., Normal, IL: Dalkey Archive Press, 1996), 339–44.
35. Barth, 342–43.
36. Barth, 343.
37. Barth, 351.
38. Review of *Sabbatical: A Romance*, by John Barth, *Kirkus Reviews*, May 1, 1982, https://www.kirkusreviews.com/book-reviews/john-barth/sabbatical/; Victor Howes, "Barth's Literary Puzzle," review of *Sabbatical: A Romance*, by John Barth, *Christian Science Monitor*, July 9, 1982.
39. John Barth, "The Art of Fiction No. 86," interview by George Plimpton, *Paris Review*, Spring 1985, 154.
40. Frizzell Interview I.
41. The Brandon reference is from Frizzell Interview I; the quote is from Frizzell Interview II.
42. Frizzell Interview II.
43. Michael Frizzell, "The Chesapeake Bay Serpent," *Crypto Special Number 1: Dracontology: Being an Examination of Unknown Aquatic Animals*, November 2001, 130.
44. Frizzell Interview II.
45. Frizzell, "Chesapeake Bay Serpent," 130.
46. Robinson, "Chessie May Have Made."
47. Siemer, "'Chessie' Sighting Claimed."
48. David Dudley, "Chessie's Dead: Ten Years after the TV Shows and the T-Shirts, Beloved Local Monster Sinks Out of Sight. Who Killed the Bay's Reluctant Sea Serpent?," *Baltimore City Paper*, June 25, 1993, 12; Siemer, "'Chessie' Sighting Claimed."
49. Chip Brown, "Creature Feature," *Washington Post*, July 16, 1982.
50. Frizzell, "Chesapeake Bay Serpent," 131.
51. Bill Burton, "Desperately Seeking Chessie," *Chesapeake Bay Magazine*, January 1995, 30.

Chapter 3. Dissecting the Frew Film

1. Mary Siemer, "'Chessie' Sighting Claimed; Serpent Said Caught on Film," *KIBT*, June 30, 1982; Karen Frew quoted in Russ Robinson, "Woman Kept Mum on Sea Creature Photos," *BES*, July 18, 1982; Russ Robinson, "Chessie May Have Made Video Debut," *BS*, July 11, 1982.

2. Robinson, "Chessie May Have Made"; Michael Frizzell, "The Chesapeake Bay Serpent," *Crypto Special Number 1: Dracontology: Being an Examination of Unknown Aquatic Animals*, November 2001, 131.
3. Robinson, "Chessie May Have Made."
4. Frizzell Interview II.
5. George Zug to Mrs. Peabody, March 3, 1981, folder 2, Cryptozoological Correspondence Miscellaneous, SIA 95-031.
6. Frizzell Interview II.
7. J. Richard Greenwell, "Chesapeake Bay Monster Filmed on Videotape," *ISC Newsletter* 1, no. 2 (Summer 1982): 9–10.
8. Albert Oetgen and Richard E. Gordon, "Witness, Scientist Seek to Be like Chessie: Out of the Public Eye," *RTD*, July 13, 1982.
9. Mary Siemer, "Chessie Hoopla Stymies Studies," *SD*, July 14, 1982.
10. UPI, "Panel to See Tapes of Alleged Monster," *Indianapolis Star*, July 13, 1982.
11. Oetgen and Gordon, "Witness, Scientist Seek."
12. Joe Perdue to George Zug, July 12, 1982, folder 1, Correspondence from the Public, 1982, SIA 12-057; George Zug to Joe Perdue, December 13, 1982, folder 2, Cryptozoological Correspondence Miscellaneous, SIA 95-031.
13. Mary Siemer, "Chessie: Couple's Bay Monster Film Reaps Plenty of Attention," *SD*, July 13, 1982; Frizzell Interview I. Frizzell's memories of the specifics are vague.
14. "The Chessie Advantage," *SD*, July 1, 1982.
15. "Psst! Did You Know This about the Shore?," *Mid-shore Guidebook '82*, July 1, 1982, 6. This is a magazine supplement in the *SD* issue of same date.
16. Siemer, "Chessie Hoopla Stymies Studies."
17. Siemer.
18. James Holechek, "A Sea Serpent? Maybe, but Most Likely Something Ordinary," *BS*, July 18, 1982.
19. On this point, and many others relating to the perils and complexities of visuality, see James Elkins, *The Object Stares Back: On the Nature of Seeing* (New York: Simon and Schuster, 1996).
20. Loren Coleman and Patrick Huyghe, *The Field Guide to Lake Monsters, Sea Serpents, and Other Mystery Denizens of the Deep* (New York: Tarcher/Penguin, 2003), 26.
21. Siemer, "'Chessie' Sighting Claimed."

22. AP, "Chessie Publicity Takes Toll," *SD*, July 22, 1982; Robinson, "Woman Kept Mum."
23. "Scientist Will Study 'Chessie' Tape," *RTD*, July 15, 1982. On the Champ seminar, including a summary of Zug's contribution that briefly references Chessie, see Joseph W. Zarzynski, *Champ, Beyond the Legend*, updated ed. (Wilton, NY: M-Z Information, 1988), 125–39.
24. "Closet witnesses" comes from Clara Germani, "The Return of Chessie and Friends," *Christian Science Monitor*, July 27, 1982. On further sightings, see Frizzell, "Chesapeake Bay Serpent," 131.
25. Frizzell, "Chesapeake Bay Serpent," 131–32; Frizzell Interview II.
26. Robinson, "Woman Kept Mum"; Frizzell, "Chesapeake Bay Serpent," 132.
27. J. Richard Greenwell, "Chessie Videotape Analysis Inconclusive," *ISC Newsletter* 2, no. 1 (Spring 1983): 9.
28. Frizzell Interview II.
29. Michael Frizzell to George Zug, August 9, 1982, folder 2, Chessie/Enigma Project Meeting, August 20, 1982, SIA 12-057.
30. George Zug, "Summer with Chessie: 1982," folder 2, Chessie/Enigma Project Meeting, August 20, 1982, SIA 12-057.
31. This text is drawn from Charles McDowell, "The Sea Monster Makes Some News," *RTD*, August 26, 1982. An apparent draft version with minor differences can be found in folder 2, Chessie/Enigma Project Meeting, August 20, 1982, SIA 12-057.
32. Peter Hardin, "'Chessie' Evidence Intrigues Scientists," *Richmond News Leader*, August 24, 1982.
33. Michael Frizzell to George Zug, August 27, 1982, folder 2, Chessie/Enigma Project Meeting, August 20, 1982, SIA 12-057.
34. Craig Phillips, letter to the editor, *ISC Newsletter* 1, no. 4 (Winter 1982): 11.
35. Nick Carter, email to the author, July 26, 2018.
36. Frizzell Interview II.
37. George Zug, undated form letter, folder 1, Correspondence from the Public, 1982, SIA 12-057.
38. Frizzell, "Chesapeake Bay Serpent," 132.
39. Russ Robinson, "Chessie Seems Real, Scientists Say; but a Real What, They Don't Know," *BES*, August 31, 1982.
40. Guy Friddell, "Chessie Monster Is a Star as Scientists Study Videotape," *Virginian-Pilot* (Norfolk, VA), October 4, 1982.

41. McDowell, "Monster Makes Some News."
42. J. Richard Greenwell to George Zug, September 3, 1982, folder 6, ISC Correspondence, 1981–1982, SIA 95-031; Zug to Greenwell, September 8, 1982, folder 6, ISC Correspondence, 1981–1982, SIA 95-031; Greenwell, "Chessie Videotape Analysis Inconclusive," 9.
43. "Chessie Update," *Champ Channels* 1, no. 3 (Wilton, NY: Lake Champlain Phenomena Investigation, December 1983); "Dr. Zug on Chessie," *Champ Channels* 2, no. 1 (Wilton, NY: Lake Champlain Phenomena Investigation, 1984).
44. Frizzell, "Chesapeake Bay Serpent," 132.
45. Frizzell, 132; Michael Frizzell to George Zug, August 27, 1982, folder 2, Chessie/Enigma Project Meeting, August 20, 1982, SIA 12-057.
46. Robinson, "Chessie Seems Real"; Michael Frizzell to George Zug, September 11, 1982, folder 2, Cryptozoological Correspondence Miscellaneous, SIA 95-031.
47. Russ Robinson, "Elusive 'Chessie' Stumps Computer," *BS*, August 5, 1984.
48. Russ Robinson, "Enhancement Permits Detailed Look at 'Chessie,'" *BS*, August 5, 1984; Robinson, "Elusive 'Chessie' Stumps Computer."
49. Robinson, "Enhancement Permits Detailed Look"; quote from David Dudley, "Chessie's Dead: Ten Years after the TV Shows and the T-Shirts, Beloved Local Monster Sinks Out of Sight. Who Killed the Bay's Reluctant Sea Serpent?," *Baltimore City Paper*, June 25, 1993, 12.
50. Frizzell, "Chesapeake Bay Serpent," 133.
51. William Rodgers, letter to the editor, *SD*, August 17, 1984.

Chapter 4. A Serpent in Eden

1. For the Potomac sighting, see UPI, "Did Captain Spot 'Chessie' or Trash?," *RTD*, May 1, 1983; on the whale, see Pat Emory, "It's a Whale of a Tale," *KIBT*, June 22, 1983.
2. Charles McDowell, "Where Is Chessie?," *RTD*, July 10, 1983.
3. Philip M. Boffey, "'Chessie' Back in the Swim Again," *NYT*, September 4, 1984.
4. Chessie, Oetgen wrote, had "apparently decided to spend 1981 in seclusion. For the discerning Chessie follower, however, that is explainable behavior. . . . The year 1981, after all, was an odd-numbered one and Chessie has demonstrated an affinity for privacy in odd-numbered

years." Albert Oetgen, "What's Happened to Elusive 'Chessie'?," *RTD*, December 27, 1981.

5. Burton advanced the bluefish theory for the first time in the aftermath of the Frew sighting. See Bill Burton, "Does Chessie Like Bluefish?," *BES*, [July?] 1982, PFPA.
6. Dan Rodricks, "Fear Not but Tread Lightly; Old Chessie Is Well-Worn," *BES*, October 29, 1982.
7. "The Great Sea Monster Mystery," presented by Fergus Keeling, BBC Radio 4, December 27, 1986.
8. AP, "Elusive Chessie Just Won't Compute," *SD*, August 6, 1984.
9. Peter Jensen, "Chessie Spotted Again," *SD*, August 7, 1984; Lisa Lister, "Jumpin Catfish . . . It's Chessie Again!," *KIBT*, August 8, 1984. A *Washington Post* article in late 1984 suggested that Harry Lohman "saw Chessie twice during a three-day period," but there is no evidence of this story elsewhere. Denis Collins, "'Chessie' and His Ilk: Demonsterably Shy," *Washington Post*, December 9, 1984.
10. Lisa Lister, "Chessie Gives Encore Appearance," *KIBT*, August 15, 1984.
11. Lister.
12. "Island View: Do You Believe in 'Chessie'?," *KIBT*, August 15, 1984.
13. "Island View."
14. Lisa Lister, "Chessie Sighted: Third Consecutive Week," *KIBT*, August 22, 1984.
15. "Observations: Stranger Things Have Happened," *KIBT*, August 22, 1984.
16. Lisa Lister, "Something about Chessie . . . ," *KIBT*, August 22, 1984.
17. Paul McKnight, "New 'Chessie Central' Center," *KIBT*, October 3, 1984.
18. McKnight.
19. Paul McKnight, "Remarkably Similar 'Monster' Stories," *KIBT*, October 3, 1984.
20. "Subscription Offer Clarification," *KIBT*, May 22, 1985. Mug art specifics are from an interview with a staffer in the early 1990s, in David Dudley, "Chessie's Dead: Ten Years after the TV Shows and the T-Shirts, Beloved Local Monster Sinks Out of Sight. Who Killed the Bay's Reluctant Sea Serpent?," *Baltimore City Paper*, June 25, 1993, 5.
21. Boyd Gibbons, *Wye Island* (Baltimore: Johns Hopkins University Press, 1977), 113.
22. George H. Callcott, *Maryland and America, 1940 to 1980* (Baltimore: Johns Hopkins University Press, 1985), 11–12.

23. Gibbons, *Wye Island*, 22.
24. "William Preston Lane Jr. Memorial (Bay) Bridge (US 50/301)," Maryland Transportation Authority, https://mdta.maryland.gov/Toll_Facilities/WPL.html.
25. Gibbons, *Wye Island*, 117–21.
26. Janet Freedman, *Kent Island: The Land That Once Was Eden* (Baltimore: Maryland Historical Society, 2002), xii.
27. For the numbers, see Paul McKnight, "Kent Island Grows and Groans," *KIBT*, June 22, 1983; "QA Co.: Fastest Growth on the Shore," *KIBT*, November 27, 1985; and Paul McKnight, "Now We Have the Figures . . . ," *KIBT*, January 1, 1986. The quote comes from Mary K. Tilghman, "Candidate Pascal Warns about Growth," *KIBT*, March 3, 1982.
28. Nathaniel T. Kenney, "Chesapeake Country," *National Geographic*, September 1964, 370–411.
29. James A. Michener, *Chesapeake* (New York: Random House, 1978), x.
30. John Barth, *Sabbatical: A Romance* (1982; repr., Normal, IL: Dalkey Archive Press, 1996), 341.
31. Greg Couteau, "Michener Entices World to Chesapeake," in "Chesapeake Country: A Report on Social and Economic Trends in the Mid-shore," supplement, *KIBT*, January 27, 1982.
32. John R. Wennersten, *Maryland's Eastern Shore: A Journey in Time and Place* (Centreville, MD: Tidewater, 1992), 268.
33. Pat Emory, "Chessie Gets a New Assignment," *SD*, October 8, 1984.
34. On the rise and fall of David Nichols, see Gibbons, *Wye Island*, 117–21.
35. "$40 Million Housing Begins," *KIBT*, November 20, 1985.
36. Wennersten, *Maryland's Eastern Shore*, 267.
37. Louis C. Timms, "Crying Waterman Is Island Symbol," letter to the editor, *SD*, January 3, 1984.
38. Paul McKnight, "Will Our Children Be Refugees from Their Own Home Towns?," *KIBT*, October 16, 1985.
39. On the relationship between wetlands protection and development, see Adam Rome, *The Bulldozer in the Countryside: Suburban Sprawl and the Rise of American Environmentalism* (New York: Cambridge University Press, 2001), 153–65.
40. On urban and suburban attitudes to the countryside and its preservation, see Samuel P. Hays, *Beauty, Health, and Permanence: Environmental Politics in the United States, 1955–1985* (New York: Cambridge University Press, 1987), 22–39. Also see Hal K. Rothman, *The Greening of a Nation?*

Environmentalism in the United States since 1945 (Philadelphia: Harcourt Brace College, 1998), 102–3.

41. Oyster photograph, *KIBT*, December 31, 1986.

Chapter 5. Creating a (State) Monster

1. Russ Robinson, "Elusive 'Chessie' Stumps Computer," *BS*, August 5, 1984.
2. "Chessie Update—1984," *Champ Channels* 2, no. 3 (Wilton, NY: Lake Champlain Phenomena Investigation, December 1984).
3. Della Interview.
4. TV listings, *Star-Democrat Weekend Magazine*, May 28, 1982.
5. Reviews of *Baby* make for painful reading. Although he was not entirely negative, Dick Fleming of the Salisbury, Maryland, *Daily Times* wrote, for example, that the film "attempts to provide fantasy entertainment that is wholesome enough for youngsters and intelligent enough for discriminating adult viewers. It succeeds only in the sense that it is better than nothing." "Legend Is Best Left Lost," *Daily Times* (Salisbury, MD), March 31, 1985.
6. Frederick Rasmussen, "So Long, 'Captain Chesapeake,'" *BS*, December 20, 2000.
7. David Dudley, "Chessie's Dead: Ten Years after the TV Shows and the T-Shirts, Beloved Local Monster Sinks Out of Sight. Who Killed the Bay's Reluctant Sea Serpent?," *Baltimore City Paper*, June 25, 1993, 10.
8. Neff Hudson, "Do They Exist? Chessie," *AC*, October 9, 1988.
9. "'Chessie's' Back," *Talbot Banner* (Easton, MD), June 8, 1983.
10. Tom Cheezum II, letter to the editor, *SD*, August 13, 1984.
11. On the early history of regional Chesapeake Bay efforts, see Howard R. Ernst, *Chesapeake Bay Blues: Science, Politics, and the Struggle to Save the Bay* (New York: Rowman and Littlefield, 2003), 11–14. On Maryland's conservation activities from founding to 1980, see George H. Callcott, *Maryland and America: 1940 to 1980* (Baltimore: Johns Hopkins University Press, 1985), 259–75.
12. Ernst, *Chesapeake Bay Blues*, 11–18.
13. On Marylanders' concern for the environment as a political interest, see Herbert C. Smith and John T. Willis, *Maryland Politics and Government: Democratic Dominance* (Lincoln: University of Nebraska Press, 2012), 100–103.

14. Syd Courson, "Potomac Sea Serpent Grants Exclusive (Mythical) Interview," *News* (Frederick, MD), September 1, 1978.
15. Jack Greer, "Bloody Pt. Hole May Become Dump," *KIBT*, January 27, 1982.
16. Thomas F. Flannery, "Unidentified Creature Reported in Bay Again . . . News Item," political cartoon, *BS*, July 18, 1982.
17. US Fish and Wildlife Service, "Are We Ready for 'Bigfoot' or the Loch Ness Monster?," news release, December 21, 1977.
18. Joseph W. Zarzynski, *Champ, Beyond the Legend*, updated ed. (Wilton, NY: M-Z Information, 1988), 112–20.
19. "Chessie Update," *Champ Channels* 1, no. 3 (Wilton, NY: Lake Champlain Phenomena Investigation, December 1983).
20. George W. Earley, "Save the Chesapeake Monster," *Baltimore News American*, April 1, 1981.
21. Michael Frizzell to George Della, September 7, 1984, SJR13-Della.
22. On tourism and Champ protection laws, see Robert E. Bartholomew, *The Untold Story of Champ: A Social History of America's Loch Ness Monster* (Albany: State University of New York Press, 2012), 113–19.
23. Michael Frizzell to George Della, September 7, 1984, SJR13-Della.
24. George Della to Joseph Bernstein, October 2, 1984, SJR13-Della.
25. Draft of Senate Joint Resolution 13, undated, SJR13-MD.
26. Michael Frizzell to George Della, October 28, 1984, SJR13-MD.
27. AP, "State Monster May Be Next," *SD*, January 21, 1985.
28. "Gallimaufry," *BS*, January 21, 1985.
29. The two Della quotes are from Della Interview; Christopher Crowe to ?, January 23, 1985, SJR13-MD.
30. Economic and Environmental Affairs Committee agenda, *BS*, January 27, 1985.
31. Michael Frizzell, "The Chesapeake Bay Serpent," *Crypto Special Number 1: Dracontology: Being an Examination of Unknown Aquatic Animals*, November 2001, 134.
32. Della Interview.
33. Maryland Department of Natural Resources, bill report, SJR 13, January 22, 1985, SJR13-MD.
34. Liz Bowie, "Senators Ask What Chessie Is Likely to Eat," *SD*, January 30, 1985.
35. The first Clark quote is from Dan Fesperman, "'Chessie' Surfaces during Annapolis Hearing," *BES*, January 30, 1985. Subsequent quotes

are from AP, "Lawmakers Are Wary of Chessie Monster Tape," *Daily Times* (Salisbury, MD), January 30, 1985.

36. Frizzell Interview I.
37. AP, "Chessie: Lawmakers Asked to Protect Creature," *Frederick (MD) Post*, January 30, 1985.
38. Anne Hazard, "More Claim to Have Seen Strange Aquatic Creatures," *RTD*, August 17, 1978.
39. Anne Hazard, "Like Serpent, Bid to Help It Goes Under," *RTD*, August 19, 1978; Potomac River Fisheries Commission meeting minutes, August 18, 1978, http://prfc.us/pdfs/19780818.pdf.
40. Albert Oetgen, "Non-monster 'Very Explainable,' Latest Chessie Observers Say," *RTD*, September 21, 1980.
41. Mary Siemer, "'Chessie' Draws Nation's Eyes to Kent Island," *KIBT*, July 14, 1982; Lisa Lister, "Chessie Gives Encore Appearance," *KIBT*, August 15, 1984.
42. "Statement of Mike Frizzell," undated [ca. January 29, 1985], SJR13-Della.
43. "Statement of Mike Frizzell."
44. Joseph W. Zarzynski to the Maryland legislature, January 25, 1985, SJR13-Della.
45. Bowie, "Senators Ask."
46. Peter Jensen, "Rockfish Ban Official Now," *SD*, January 2, 1985. The Clark quotes are from Fesperman, "'Chessie' Surfaces."
47. AP, "Chessie: Lawmakers Asked."
48. Tom White, "Chessie Deserves Her Place in the Bay as State's Official Sea Serpent," *Baltimore News American*, February 1, 1985.
49. "You Said It," *AC*, February 6, 1985.
50. "Chessie?," *St. Louis Post-Dispatch*, February 4, 1985.
51. SJR13 voting record, February 6, 1985, SJR13-MD.
52. "Chessie: Senate Won't Admit Bay Creature May Exist," *BS*, February 7, 1985.
53. Della Interview.
54. Frizzell Interview I.
55. Dudley, "Chessie's Dead," 16.
56. "Chessie Monster Gets No Respect," *AC*, February 7, 1985.
57. Joseph W. Zarzynski, letter to the editor, *BS*, March 16, 1985.

Chapter 6. A Chesapeake Bay Story

1. Kinser Interview I.
2. "Glenn Kinser: Letting Nature Take Care of Nature," *Shepherdstown (WV) Chronicle*, July 16, 2010; "Glenn William Kinser, Jr. (1940–2013)," *FWS Retirees Association Newsletter*, Fall 2013, https://www.fwsretirees.org/Newsletters/Fall_2013.pdf.
3. Glenn Kinser, email to the author, January 31, 2005; Kinser Interview I.
4. Kinser, email to the author.
5. Kinser Interview I.
6. Glenn Kinser, email to the author, January 31, 2005.
7. Glenn Kinser, email to the author, April 19, 2005; Kinser Interview I.
8. Kinser Interview I.
9. On charismatic megafauna and efforts to protect it, see Shannon Petersen, *Acting for Endangered Species: The Statutory Ark* (Lawrence: University Press of Kansas, 2002), 21-35, 41, 58. Chessie's story especially resembles that of dolphins and porpoises, which have become representations of the ocean and its environmental problems for many Americans. See Gregg Mitman, *Reel Nature: America's Romance with Wildlife on Film* (Cambridge, MA: Harvard University Press, 1999), 157–79.
10. Kinser Interview I.
11. Harald Fuller-Bennett and Iris Velez, "Woodsy Owl at 40," *Forest History Today*, Spring 2012, 22–27.
12. Eve Horney, "Chessie Spotted with Companion," *KIBT*, March 13, 1985.
13. Horney.
14. See, for instance, the advertisement on page 35 of *AC*, November 21, 1985.
15. Bill Burton, "Chessie Resurfaces, but No Still Photo Yet," *BS*, May 5, 1985.
16. Christine Reid, "Bolder, Bigger Chessie Espied in Potomac," *RTD*, July 12, 1985.
17. Charles McDowell, "On Looking for Chessie," *RTD*, July 16, 1985.
18. Kinser Interview I.
19. US Fish and Wildlife Service, *Wetlands Coloring Book*, 2, with handwritten and typed comments, undated, GKPA.
20. US Fish and Wildlife Service, *Nature Series: Fisheries and Me*, 7, with handwritten and typed comments, undated, GKPA.
21. "Coloring Book—Guidelines for Artist," undated, GKPA.
22. "Guidelines for Artists," undated, GKPA.

23. Marianne Lydecker and Charlie Rewa, Chessie art samples, undated, GKPA.
24. Jamie Harms to Charlie Rewa, August 9, 1985, GKPA.
25. Kinser Interview I.
26. D. W. Woodard to Inez Connor, April 11, 1986, GKPA; Inez Connor to D. W. Woodard, May 12, 1986, GKPA.
27. Notes on coloring book guidelines, undated, GKPA.
28. "Chesapeake Bay Coloring Book—Draft Text/Graphics Ideas," undated, GKPA.
29. "Chesapeake Bay Coloring Book, Another Approach, 10 Pages," undated, GKPA.
30. Mockup #1, undated, GKPA. Emphasis in original.
31. Jamie Harms and Dave Folker, *Chessie: A Chesapeake Bay Story* (Annapolis: US Fish and Wildlife Service, 1986), 6 and front cover. The suggested artwork in the book's early draft was for a "friendly looking picture of Chessie in the water[,] perhaps in front of the Bay Bridge or another familiar landmark." "Chesapeake Bay Coloring Book—Draft Text/Graphics Ideas," undated, GKPA.
32. Harms and Folker, *Chessie*, 8. On omission of Virginia from the map, see Kinser Interview I.
33. Glenn Kinser, email to the author, January 31, 2005.
34. D. W. Woodard to Inez Connor, April 11, 1986, GKPA.
35. Inez Connor to D. W. Woodard, May 12, 1986, GKPA.
36. Albert d'Amato to Inez Connor, July 16, 1986, GKPA.
37. Jamie Harms to Inez Connor, July 22, 1986, GKPA.
38. Glenn Kinser, email to the author, January 31, 2005.
39. Soraya Sarhaddi, "Retiree Reports Seeing Chessie," *SD*, March 10, 1986.
40. Jeff McCulley, "Did Ol' Chessie Surface on the Tred Avon?," *SD*, May 28, 1986.
41. Details of the Bishop-Boudrie sighting and its aftermath are drawn from the following news articles: Suzanne E. Murray, "Sighting Reported on Tred Avon River: Chessie—Is That You?," *KIBT*, June 4, 1986; McCulley, "Did Ol' Chessie Surface?"; David Dudley, "Chessie's Dead: Ten Years after the TV Shows and the T-Shirts, Beloved Local Monster Sinks Out of Sight. Who Killed the Bay's Reluctant Sea Serpent?," *Baltimore City Paper*, June 25, 1993. The "starship," "many," and "reliable business people" quotes are from Rex Springston, "Chessie Is Still Splashing Around," *Richmond News Leader*, July 24,

1986. On the Snead encounter, see Michael Frizzell, "The Chesapeake Bay Serpent," *Crypto Special Number 1: Dracontology: Being an Examination of Unknown Aquatic Animals*, November 2001, 129–37. Bishop's radio appearance took place on "The Great Sea Monster Mystery," presented by Fergus Keeling, BBC Radio 4, December 27, 1986.

42. Suzanne E. Murray, "KI Resident Has Chessie Explanation," *KIBT*, June 11, 1986.
43. Springston, "Chessie Is Still Splashing."
44. Effie Cottman, "Chessie Joins Cleanup Effort," *AC*, October 6, 1986.
45. Steve Funderburk to Dave Folker, November 3, 1986; Steve Funderburk to Jamie Harms, November 3, 1986; Donna L. Stotts to P. O. Baker, June 1, 1987; and letter to educators and questionnaire, undated, all in GKPA.
46. Kinser Interview I.
47. "Draft Agenda for February 20 Press Conference," February 13, 1987, GKPA.
48. "Monstrous Doings in Chesapeake Bay," *NYT*, April 19, 1987.
49. Slattery Interview.
50. Kinser Interview I.

Chapter 7. Diminishing Returns

1. Foer Interview.
2. Foer Interview.
3. Foer Interview.
4. "Search for Chessie," *AC*, August 28, 1986.
5. Judi Perlman, "Chessie: Expedition Setting Off to 'Find' Bay Creature," *AC*, August 30, 1986.
6. Except where otherwise noted, details of the 1986 expedition are drawn from the following: Nancy Hathaway Noyes, "Chasing Chesapeake's Chimera," *Publick Enterprise* (Annapolis, MD), October 1986 (first half); Suzanne E. Murray, "Elusive . . . Chessie Remains a Mystery Despite Search Expedition," *KIBT*, September 15, 1986.
7. Kenneth Chang, "Gilbert V. Levin, Who Said He Found Signs of Life on Mars, Dies at 97," *NYT*, August 4, 2021.
8. Foer Interview.
9. Noyes, "Chasing Chesapeake's Chimera"; Foer Interview.
10. Noyes, "Chasing Chesapeake's Chimera."

11. "Chessie Proves Elusive," *AC*, September 12, 1986; Noyes, "Chasing Chesapeake's Chimera."
12. Murray, "Elusive . . . Chessie Remains a Mystery Despite Search Expedition."
13. Paul Foer to Rod Coggin, September 24, 1986, PFPA.
14. Except where otherwise noted, details of the 1987 expedition are drawn from the following: Karina Paape, "Foer Prepares Another Hunt for Elusive Chessie," *AC*, August 27, 1987; "Bay Adventure Seeks 'Chessie': Search Benefits Cleanup," *AC*, September 17, 1987; Jackie Powder, "Chasing Chessie Hints at Complexity of Life in the Bay," *BS*, September 21, 1987; Kyra Scarton, "Can Chessie Save the Bay? Monster Used to Aid in Cleanup Efforts," *AC*, September 21, 1987; Elizabeth Rothschild, "Bay Science Expedition a Success Despite No Sightings of Creature," *KIBT*, September 23, 1987; Betty Duty, "MWA Goes in Search of Legendary Chessie," *Waterman's Gazette* (Annapolis, MD), November 1987.
15. Official program, September 20, 1987, PFPA.
16. Official program.
17. Foer Interview.
18. Bert Brun, "Chessie—We Need You," in official program, September 20, 1987, PFPA.
19. Paul Foer to Michael Frizzell, July 28, 1987, PFPA.
20. Michael Frizzell to Paul Foer, August 20, 1987, PFPA.
21. Official program, September 20, 1987, PFPA.
22. Voice of America broadcast, September 1987, PFPA, transcript in author's possession.
23. Scarton, "Can Chessie Save the Bay?"
24. Both quotes are from Rothschild, "Bay Science Expedition."
25. Larry Simns to Paul Foer, October 5, 1987, PFPA.
26. Rex Springston, "Separating Monsters from Myth," *Richmond News Leader*, May 16, 1988.
27. Gail Dean, "Couple, Grandson Report Seeing Chessie in Brannock Bay Waters," *SD*, May 18, 1988.
28. Wilford Kale, "Chessie Sightings Good News, Says Loch Ness Sleuth," *RTD*, May 22, 1988.
29. Neff Hudson, "Do They Exist? Chessie," *AC*, October 9, 1988.
30. "Chesapeake Corner: Chessie Still a Mystery," *AC*, April 16, 1987; "Chesapeake Corner: Bay Educational Resources Are Easy to Find," *AC*, December 17, 1987.

31. Slattery Interview.
32. Britt Eckhardt Slattery and Dave Folker, *Chessie Returns!* (1988; repr., Annapolis, MD: US Fish and Wildlife Service Chesapeake Bay Field Office, 1993), 9–10.
33. Kinser Interview I.
34. Slattery Interview.
35. Tim Sayles, "The Great Chessie Chase: Who Knows What Mysterious Creatures Lurk beneath the Waters of the Chesapeake," *Mid-Atlantic Country*, September 1989, 80.
36. "Bay News," ca. 1990, GKPA.
37. US Fish and Wildlife Service Chesapeake Bay Estuary Program, "First Progress Report," October 1990; US Fish and Wildlife Service Chesapeake Bay Estuary Program, "1991 Annual Report."
38. Slattery Interview.
39. Irene J. Solovij, "From Songs to Books, This Woman Bay-Inspired," *AC*, May 13, 1988.
40. Holland Interview.
41. Mary Carole McCauley, "Kids' Author Ventures into Adult Fiction," *BS*, December 9, 2012.
42. Margaret Meacham, *The Secret of Heron Creek* (Centreville, MD: Tidewater, 1991), 16.

Chapter 8. Stand by Your Manatee

1. Bill Burton, "Desperately Seeking Chessie," *Chesapeake Bay Magazine*, January 1995, 30.
2. Burton, "Desperately Seeking Chessie."
3. AP, "Manatee Seen Near Rock Hall," *News* (Frederick, MD), September 10, 1994.
4. Leslie Gross, "Manatee's New Home: Kent Narrows?," *AC*, September 20, 1994.
5. Virginia experienced confirmed manatee visits in both 1985 (see AP, "Tropical Sea Mammal Strays into VA. Waters," *RTD*, September 6, 1985; and AP, "Manatee Is Eluding Rescuers," *RTD*, September 7, 1985) and 1987 (see Rex Springston, "'It Moved the Boat': Manatee Sighting May Be Credible Fish Story," *RTD*, August 18, 1987; LeeNora Everett, "Huge Creature Is Thought to Be Manatee," *RTD*, August 19, 1987; and Bill Geroux, "'Manatee' Misses Breakfast," *RTD*, August 20, 1987).

Another visit occurred in September 1992 (see "Manatee Sighted by Men at Deep Creek Lock," *RTD*, September 30, 1992).

6. Geroux, "'Manatee' Misses Breakfast."
7. "Search for Manatee Continues," *AC*, September 26, 1994; Michael Cody, "As Water Cools, Hunt for Manatee Heats Up," *AC*, September 28, 1994. The Lehr quote is from the latter article.
8. "Manatee Released in Florida," *AC*, October 9, 1994. Funding sources for the operation are reported in AP and Christopher Munsey, "Endangered Manatee Finally Corralled," *AC*, October 2, 1994.
9. "Endangered Manatee Finally Corralled"; "Manatee Enthusiasts Open Sunday 1–4," *AC*, October 9, 1994.
10. Bangert Interview.
11. "Hail and Farewell to Our Manatee," *AC*, October 10, 1994.
12. "Frisky" comes from Leslie Crook, "Chessie Missing Again," *SD*, November 2, 1994. Also see "Jabez Land Should Be Saved," *BS*, November 9, 1994; and Christopher Munsey, "A Blaze of Holiday Light," *AC*, December 9, 1994.
13. Quoted in Tom Horton, "Unforgettable Gifts from Nature," *BS*, December 24, 1994.
14. Crook, "Chessie Missing Again."
15. AP, "Rescued Manatee Returns," *AC*, July 8, 1995.
16. On Chessie's new tracking system, see Edward Lee, "Vising Manatee Favors Other Haunts This Year," *BS*, July 19, 1995; Dale Hopper, "Wandering Manatee Monitored from Space," *News Journal* (Wilmington, DE), July 21, 1995; and Leslie Crook, "Chessie May Be Looking for His Roots," *SD*, July 21, 1995. On Chessie's route, see AP, "Wayward Manatee Makes Way toward Atlantic Ocean," *AC*, July 19, 1995; and AP, "Wayward Manatee Seen Near Assateague Island," *AC*, July 20, 1995.
17. AP, "Rescued Manatee Returns."
18. Robert Perkins, "Beware the Manatee Precedent," *SD*, July 16, 1995.
19. "He Likes Our Grass," *SD*, July 23, 1995 (editorial reprinted from *Kent County News*).
20. Michael Humphreys, letter to the editor, *News Journal* (Wilmington, DE), July 31, 1995.
21. Craig Quintana, "Federal Budget Cuts Force Florida Researchers to Curtail Manatee Project," *Washington Post*, November 24, 1995; Brad Knickerbocker, "Endangered Species Act Faces Its Own Dangers," *Christian Science Monitor*, March 8, 1995.

22. "Manatee Appreciates Area's Hospitality," *AC*, July 20, 1995.
23. Anthony G. Girandola Jr., letter to the editor, *BS*, October 14, 1994.
24. "Where's the Outrage?" *SD*, October 27, 1994.
25. "Manatee 'Browsing' in Delaware Bay," *BS*, July 30, 1995.
26. Leslie Crook, "Atlantic City Glitz Beckons to Chessie," *SD*, August 1, 1995.
27. John Curran, "A Wayfaring Manatee Broadens His Horizons," *Courier-Post* (Camden, NJ), August 3, 1995.
28. Todd B. Bates, "Sea Cow? Sure Did," *Asbury Park (NJ) Press*, August 5, 1995.
29. Chuck Sudetic, "On the Trail of a Vagabond Manatee," *NYT*, August 8, 1995.
30. Michael Daly, "His Love Hunt's Only Human," *Daily News* (New York, NY), August 13, 1995.
31. Sudetic, "On the Trail."
32. Mickey Hackman, "Chessie the Wandering Manatee Waves Farewell to New York Harbor," *Daily News* (New York, NY), August 10, 1995.
33. Mark Pazniokas, "'Sea Monster' Moseys into State Waters," *Hartford Courant*, August 16, 1995; Brigitte Greenberg, "Wayward Manatee in L.I. Sound," *Record-Journal* (Meriden, CT), August 16, 1995.
34. Denis Horgan, "The Gentle, Lovable Manatee Is Just What State Needs," *Hartford Courant*, August 18, 1995.
35. Mark Pazniokas, "Manatee Spurns State for Rhode Island Waters," *Hartford Courant*, August 18, 1995; Robert Braile, "Stray Manatee Nears Cape," *Boston Globe*, August 20, 1995. The "tourist" quote is from the first article.
36. Mark Pazniokas, "Chessie's Hope: Go South, Young Manatee," *Hartford Courant*, August 22, 1995.
37. Mark Pazniokas, "Wayward Manatee Stops Phoning Home," *Hartford Courant*, August 25, 1995; AP, "Scientists Lose Chessie Off Coast of Connecticut," *RTD*, August 25, 1995.
38. Brigitte Greenberg, "Biologists Don't Always Get Their Manatee," *Bennington (VT) Banner*, August 26, 1995.
39. John A. Harnes, "Wayward Manatee on Way Home," *Asbury Park (NJ) Press*, September 13, 1995.
40. AP, "'Chessie' Spotted in Lower Chesapeake," *News* (Frederick, MD), September 22, 1995; Leslie Crook, "Briefly Feared Dead, Chessie Headed Home," *SD*, September 28, 1995.
41. AP, "Chessie's Back in Fla.," *SD*, November 22, 1995.

42. Michael Daly, "He's Man(atee) of the Year," *Daily News* (New York, NY), October 29, 1995.
43. "Chessie's Back Home," *Hartford Courant*, November 28, 1995.
44. John A. Harnes, "Fla. Manatee Returns Home for Holidays," *Asbury Park (NJ) Press*, November 22, 1995.
45. Stephen Barr, Thomas W. Lippmann, and Bill McAllister, "The 1996 Budget: Winners and Losers," *Washington Post*, April 29, 1996.
46. Leslie Crook, "Biologists Catch Up with Chessie," *SD*, February 23, 1996; Leslie Crook, "Chessie Up for Adoption," *SD*, February 23, 1996.
47. "Chessie Gets Outfitted with New Transmitter," *Daily Times* (Salisbury, MD), March 13, 1996.
48. Leslie Crook, "Road North Beckons Once Again to Chessie," *SD*, June 17, 1996; "Chessie the Manatee Moving North toward Md. Again," *BS*, July 4, 1996.
49. "Manatee Is Cruising Coastal North Carolina," *BS*, July 12, 1996. Information about Hurricane Bertha comes from "Hurricane Bertha," National Weather Service, https://www.weather.gov/mhx/Jul121996EventReview.
50. Leslie Crook, "Chessie Ditches His Transmitter," *SD*, July 19, 1996.
51. On possible Maryland sightings, see Mary Ellen Lloyd, "Chessie May Be in Pasadena," *AC*, July 31, 1996; and AP, "Chessie the Manatee Seen Near Annapolis," *Daily Times* (Salisbury, MD), August 13, 1996. In Virginia, see Rex Springston, "Traveling Manatee Loses Baggage in Waterway," *RTD*, July 20, 1996; and AP, "Manatee Believed Sighted in State," *RTD*, August 23, 1996.
52. Leslie Crook, "As Chessie Heads to Florida, Bay Still Attracts Manatees," *SD*, September 5, 1996.
53. Ellen Gamerman, "That Ripple Probably Wasn't Chessie," *BS*, August 15, 1996.
54. "Editor's Notebook," *AC*, August 17, 1996; Dan Berger, "On the Other Hand . . . ," *BS*, August 19, 1996.
55. Angela Price, "Chessie Believed to Have Spent Time in Chesapeake," *SD*, September 30, 2001.
56. Steve Kilar and Timothy B. Wheeler, "Chessie the Manatee Pays Return Visit to Maryland," *BS*, July 16, 2011.
57. "Chessie: Chesapeake by Choice," *BS*, July 19, 2011.
58. Price, "Chessie Believed."
59. US Fish and Wildlife Service Customer Service Center, email to the author, January 27, 2005.

Chapter 9. Familiar Friend

1. George J. Collins and Kathy Alexander, *Chessie Racing: The Story of Maryland's Entry in the 1997–1998 Whitbread Round the World Race* (Baltimore: Johns Hopkins University Press, 2001), 1.
2. Collins and Alexander, 12.
3. Collins and Alexander, 35.
4. Collins and Alexander, 13–15.
5. Collins and Alexander, 175.
6. Collins and Alexander, 179.
7. Collins and Alexander, 181, 187.
8. Collins and Alexander, 209. After the race, the boat was sold, repainted, and used to train sailors for a 2001–2 race. Proceeds went to Living Classrooms.
9. Exotic Paddle Boats USA homepage, now defunct (copy in author's possession).
10. "Preakness Fun," *Frederick (MD) Post*, May 11, 2001.
11. Andrea Cecil, "Couple Forms New Business Building Fiberglass Paddleboats," *Baltimore Daily Record*, May 23, 2003.
12. Cecil.
13. "Dorchester Unveils New Branding Campaign," *SD*, October 20, 2013; "Water Moves Us in Dorchester County," Dorchester County Economic Development, https://choosedorchester.org/water-moves-us/.
14. Lee Dravis, *Key Monster* (Baltimore: America House, 2000), 10-11.
15. Dravis, 12.
16. Dravis, 10.
17. Lisa Cole, *Searching for Chessie of the Chesapeake Bay* (Sarasota, FL: Peppertree, 2011), 71.
18. Bianca Schulze, "Author Showcase: New Children's Book 'Searching for Chessie of the Chesapeake Bay,'" *Children's Book Review*, July 22, 2011, https://www.thechildrensbookreview.com/2011/07/author-showcase-new-childrens-book-searching-for-chessie-of-the-chesapeake-bay/.
19. Cole, *Searching for Chessie*, 23.
20. Cole, 35, 37.
21. Cole, 39.
22. Cole, 88, 91.
23. Cole, 105–6.

24. Lorraine Mirabella, "Ripley's Believe It or Not! Gets Lease," *BS*, February 29, 2012.
25. Lorraine Mirabella, "'Chessie' at Harbor: Believe It or Not?," *BS*, October 27, 2011.
26. Mirabella.
27. "About," RAR Brewing, https://rarbrewing.com/about/.
28. Sarah Hainesworth, "Children's Festival Has Crofton Connection," *Crofton-West County Gazette* (Crofton, MD), September 26, 2014.
29. "Bourbon Barrel-Aged Chessie," Union Craft Brewing, https://unioncraftbrewing.com/project/bourbon-barrel-aged-chessie/.
30. David Dudley, "Chessie's Dead: Ten Years after the TV Shows and the T-Shirts, Beloved Local Monster Sinks Out of Sight. Who Killed the Bay's Reluctant Sea Serpent?," *Baltimore City Paper*, June 25, 1993, 16.
31. "Chessie Survives to Swim Another Day," *SD*, February 11, 2022, 18.
32. Anne Tate, "The Chesapeake Bay Is Closer than Ever to Becoming a National Recreation Area," *Washingtonian*, July 14, 2021, https://www.washingtonian.com/2021/07/14/chesapeake-bay-national-recreation-area/.
33. Dudley, "Chessie's Dead," 16.

INDEX